Naowarat Cheeptham
Editor

Cave Microbiomes:
A Novel Resource
for Drug Discovery

Springer

Editor
Naowarat Cheeptham
Thompson Rivers University
Kamloops, BC, Canada

ISSN 2191-5385   ISSN 2191-5393 (electronic)
ISBN 978-1-4614-5205-8   ISBN 978-1-4614-5206-5 (eBook)
DOI 10.1007/978-1-4614-5206-5
Springer New York Heidelberg Dordrecht London

Library of Congress Control Number: 2012948334

Printed on acid-free paper

Springer is part of Springer Science+Business Media (www.springer.com)

*In loving memory of Boontham
Pongpleum and Diana Irene Dobson.
For Sub and Ubon Cheeptham,
and Gary G. Dobson. To Joe and Ryder
Dobson, my rocks, with love.*

−N.C.

For Airidas and Ieva.

−M.d.L.N.E.D.

About the Authors

Dr. Naowarat Cheeptham (ncheeptham@tru.ca) has been fascinated with the world of biology since she was out catching butterflies as a child with her father in her native country, Thailand. Her interest in microbiology developed while studying at Chiang Mai University (Thailand) and Hokkaido University (Japan). Since her doctoral work, Dr. Cheeptham has been interested in discovering new drugs that could be derived from rare microorganisms that thrive in extreme habitats such as caves. Currently, she is exploring the photochemotherapeutic potential of bioactive compounds produced by volcanic cave actinomycetes. Dr. Cheeptham's work was featured on Global TV and the Knowledge Network in Canada. Besides her research interests in cave microbiology, she is also drawn to pedagogical issues in microbiology education. In 2009, she was selected as one of the biology research residency scholars in the ASM/NSF Biology Research Residency Scholars Program and participated in an NSF-sponsored residency in Washington DC. She is an assistant professor at the Department of Biological Sciences, Thompson Rivers University, Kamloops, British Columbia, Canada.

Dr. Diana E. Northup (dnorthup@unm.edu) has been studying things that live in caves since 1984. She and her colleagues on the SLIME (Subsurface Life in Mineral Environments) Team are investigating how microbes help form the colorful ferro-manganese deposits that coat the walls of Lechuguilla and Spider Cave in Carlsbad Caverns National Park; how these deposits compare to surface desert/rock varnish coatings; how microbes participate in the precipitation of calcium carbonate formations called pool fingers; and the microbial diversity located in the hydrogen sulfide cave, Cueva de las Sardinas in Tabasco, Mexico, using molecular, microbiological, and microscopy techniques. Dr. Northup has been honored by having her work featured on NOVA, the BBC, National Geographic, and the Discovery Channel. She is professor emerita in the University Libraries and a visiting associate professor of Biology at the University of New Mexico, USA.

Dr. Maria de Lurdes N. Enes Dapkevicius (mariaenes@uac.pt) is an assistant professor of Microbiology at the University of The Azores. She obtained her PhD in Biotechnology, Food Technology and Nutrition in 2002 at Wageningen University (The Netherlands). Since 1987, she has been studying bacterial communities in Azorean environments. Traditional foods and the microorganisms they harbor was her first research interest. In 2004, she started carrying out research on the bacterial biofilms that are a main feature on the walls of Azorean lava tubes. Microbiological and molecular biological studies on cave-wall bacteria and their potential biotechnological uses are, presently, her main research area.

Contents

Contributors

Cristina Riquelme Gabriel, Ph.D. CITA-A, Departamento de Ciências Agrárias, Universidade dos Açores, Angra do Heroísmo, Portugal

Lory O. Henderson Department of Biology, University of New Mexico, Albuquerque, USA

Elizabeth Terese Montano Department of Biology, University of New Mexico, Albuquerque, USA

Cesareo Saiz-Jimenez, Ph.D. Instituto de Recursos Naturales y Agrobiologia, Consejo Superior de Investigaciones Cientificas (IRNAS-CSIC), Sevilla, Spain

Phil Whitfield British Columbia Speleological Federation, Kamloops, BC, Canada

Chapter 1
Advances and Challenges in Studying Cave Microbial Diversity

Naowarat Cheeptham

Introduction

In the last decade, we have seen significant changes in how we study composition and diversity of microbial communities in various environmental samples. Advances in culture-independent molecular phylogenetic techniques have made studies on microbial communities in diverse environments more attractive and feasible (Muyzer et al. 1993; Pace 1997; Torsvik et al. 1998; Hill et al. 2000; Giraffa and Neviani 2001; Kirk et al. 2004; Barton et al. 2004; Leckie 2005; Barton et al. 2006; Malik et al. 2008; Maukonen and Saarela 2009; Hirsch et al. 2010; and Northup et al. 2011). However, using modern molecular techniques alone to study both known and unknown microbial populations in an environment has its own limitations. Several studies suggest that using a combination of both culture-independent and culture-dependent methods gives a more realistic representation of the indigenous microbial diversity (Hill et al. 2000; Gurtner et al. 2000). For example in a 2000 study Gurtner and colleagues reported using of both classical cultivation techniques and molecular approaches to compare bacterial diversity on two medieval biodeteriorated wall paintings from two churches in Austria and Germany. They obtained 70 microbial sequences of 16S rDNA sequence belonging to several genera of bacteria. The molecular approach evaluated the bacterial community by Denaturing gradient gel electrophoresis (DGGE, one of the genetic fingerprinting tools), construction of 16S rDNA clone libraries, and sequence analysis of those libraries. In the same study, isolation of heterotrophic bacteria from one of the samples using Trypticase Soy Broth (TSB) agar supplemented with 10% sodium chloride (with 3 weeks of incubation at 28°C) was also done in parallel to the above-mentioned molecular approach (Heyrman et al. 1999). The isolated strains were then characterized using fatty acid methyl ester (FAME) analysis and major FAME clusters found to

N. Cheeptham (✉)
Thompson Rivers University, Kamloops, BC, Canada
e-mail: ncheeptham@tru.ca

N. Cheeptham (ed.), *Cave Microbiomes: A Novel Resource for Drug Discovery*,
SpringerBriefs in Microbiology 1, DOI 10.1007/978-1-4614-5206-5_1,
© Naowarat Cheeptham 2013

belong to the genus *Bacillus*. Results from these two approaches failed to cross-detect similar microbial flora. In the molecular approach, 70 members of *Actinobacteria* and *Proteobacteria* including *Actinobiospora, Amycolata, Halomonas, Deleya, Rhizobium,* and *Salmonella* were identified, while it is important to note that there was no *Bacillus* detected by the molecular approach. Their findings demonstrate that the combined approach of molecular and culturing techniques may provide a better understanding of the community being evaluated. There are other review and original research works that supported using a combined approach to study microbial diversity and function in a community (Dunbar et al 1999; Torsvik and Øvreås 2002; Crecchio et al 2004). Since no individual approach is completely effective to evaluate the microbial biodiversity of a given environment, this integrated approach may provide a closer representation of the microbial community. In its own context each of these approaches should be used and evaluated accordingly.

In the field of cave microbiology, studies of cave microbial diversity have been at least attempted since 1940s (Höeg 1946; Caumartin 1963; Barton and Northup 2007; Palmer 2007). However, those studies could not truly represent a comprehensive analysis of the resident status and activities of microorganisms in cave environments due to limitations of the methods used (Barton and Northup 2007). Environments in caves are unique and illustrate difficulties for microbial community investigations. These caverns are nutrient limited, of low biomass, and high in metal contents due to diverse mineral compositions from the immediate substrate (Barton 2006; Barton et al. 2006). In these original studies, culturing techniques were solely relied upon to determine the constituent microflora. Through culturing, it is generally now understood that less than 0.1% of soil microbial community were likely identifiable in such studies (Amann et al. 1995: Hill et al. 2000). It is also now understood that the roles of microbial flora include the synthesis of cave formation, through biomineralization, bioprecipitation, and speleothem (secondary mineral deposits) construction. The contribution of these subterranean microbes on the formation/degradation of caves was not observed using culturing techniques and has not been well understood until recently (Groth et al. 1999; Groth et al. 2001; Engel et al. 2001; Northup and Lavoie 2001; Barton et al. 2001; Barton and Luiszer 2005; Gonzalez et al. 2006; Barton et al 2007; Spear et al. 2007; Engel 2010; Jones 2010; Lavoie et al. 2010; Onac and Forti 2011). With a better understanding of the microflora in cave environments, there has been a gradual shift from basic questions identifying the cavern microflora to more complex ecological and community questions that evaluate their contributions in biological and geological chemistry of the cave, ecological role in the cave biomes, and special adaptations necessary for survival of the constituent species. Because of their often unique biochemistry, these organisms have been the target to identify new drugs including antibiotics, and are used as models to evaluate the possibility of life forms on other planets (Boston et al. 2001; Nakaew et al. 2009a; Nakaew et al. 2009b; Yücel and Yamaç 2010; Northup et al. 2011).

In this section, techniques, specific challenges, and suggested protocols to evaluate cave microbe flora are identified.

General Synopsis of Cave Habitats

Caves, in a geomicrobiological aspect, are nutrient-limited habitats with a variety of redox interfaces that stem from interactions between microbial activities and minerals in such environments (Northup and Lavoie 2001). Simply, there are three zones in caves depending on the amount of light, entrance, twilight (threshold), and dark zones (Barton and Northup 2007), while Howarth suggested that caves are strictly zonal and suggested five terrestrial zones according to the classification based on transitions between surface and subsurface: (1) entrance, (2) twilight, (3) transition, (4) deep, and (5) stagnant air (Howarth 1983). The entrance zone is where the surface and underground habitats meet, while the twilight zone is where the boundary of plant life to the limit of light expands across. The transition zone is in total darkness with nocturnal desiccating winds whereas areas with long-term presence of moisture and saturated atmosphere are called the deep zone. The stagnant air zone starts beyond the deep zone and exchanges air with surface slowly (Howarth 2004). There are many types of caves according to different classifications (hypogenic/epigenic, anchialine, ice, sea, karst, lavatube caves). Caves that have been studied to date are often high in humidity (90–100%) while the temperature inside stays relatively stable at low temperature with limited temperature fluctuation during the seasons (Groth et al. 1999; Engel et al. 2001; Zhou et al. 2007; Schabereiter-Gurtner et al. 2003). Recent evaluations of a warm temperature cave in Naica, Mexico, Cueva de Los Cristales caves (Crystal Caves), have found the cave to have a more constant high temperature of 44°C (112°F) throughout the year (Onac and Forti 2011).

Other general defining physical constraints in cave environments are that they have high concentrations of minerals but are generally limited with bioavailable nutrients such as organic carbon and nitrogen sources. This is not universally true as there are caves with animals such as bats or cave swifts that alternatively have very high levels of organic carbon and nitrogen sources (Bonacci et al. 2009). Notably these caves can be very high in phosphates (from guano), and the phosphate-rich cave soil/sediment are still being mined from caves around the world to be utilized as a natural fertilizer (Onac and Forti 2011). In some cave systems, notably those with surface waters or those with animals that are able to enter and leave caves, there may be sufficient surface-derived organic matters to fuel microbial communities that rival the complexity of the ones outside of caves (Engel 2010).

Generally, however, caves have low biotic potential. Caves with limited external influence are more typically to have levels of organic carbon in the 0.5 mg organic carbon (TOC) per liter which compares to a more typical level outside of caves which normally is more than 15 mg of TOC per liter (Barton and Jurado 2007). These organic levels in caves are characterized as "extreme/near-starvation" which for oligotrophic microbial habitats is levels defined as those below 2 mg of TOC per liter (Barton and Jurado 2007). Thus, any microorganisms that can survive at these levels are classified as Oligotrophs (Koch 2001). Correspondingly, oligotrophic organisms are slow-growing organisms as in their typical growth. Subsequently, organisms in these environments typically result in low population densities and typically serve as the nutrient source for higher trophic levels. In one of Barton and colleagues' work, total cell concentration had been estimated in a cave system to approximately 10^6 cells cm^{-3} (Barton et al. 2006). This compares to soils from more organically rich sources using direct fluorescence microscopic cell counts with acridine orange of 1.5×10^{10} total bacterial cells per gram (dry forest soil) and direct plate counts 4.3×10^7 cells per gram on Thornton agar with 10% soil extract (Torsvik et al. 1990). Therefore, microbial mats and slimes, full of fungi and aerobic bacteria, are locations where arthropods, when present, are found to feed in caves. These sites are significant locations for nutrient recycling, particularly nitrogen (Dickson 1979; Howarth 1983; Northup et al 2004; Cardoso 2012). In such diverse ecosystems, microorganisms from these locations can sustain and provide enough energy for higher trophic organisms. With low to no light condition caves have limited access to nutrients and hence low biomass. Studies conducted in two different cave systems suggested that sulfur-based microbial populations such as chemoautotrophs serve as the primary producers in cave environments with no outside influences able to generate sufficient energy to sustain complex ecosystems (Engel et al. 2001; Sărbu 1991; Sărbu and Popa 1992; Sărbu et al. 1996; Northup et al. 2011). These findings cement the ideas that cave dwelling organisms not only can rely on organic carbon entering cave ecosystems from the outside, but some can also utilize energy source found in the caves. The latter gives rise to chemoautotrophy way of making end meet in such harsh environment. In a few caves including the Movile cave in Romania, microorganisms derive energy from sulfur and iron minerals (Sărbu 1991; Sărbu and Popa 1992; Sărbu et al. 1996). Majority of the organic matter found in caves with access to surface photosynthetic materials is from streams, groundwater, percolating surface water and materials coming into entrances through animate or inanimate activities (Engel et al. 2001; Engel 2010). Culver's findings from his 1985 trophic relationships in an aquatic cave were noted as follows: "Because of the absence of sunlight precludes photosynthetic primary production in caves, cave stream food webs rely exclusively on particulate detritus or dissolved organic matter originating from the surface" (Simon et al. 2003). Organisms that can survive and thrive in caves are believed to have adapted and evolved to specific selective pressures presented in caves (Snider et al. 2009; Engel 2010).

Caves are diverse environments differing in both their physical and chemical compositions. Caves are commonly classified based on (1) the solid rock that caves developed within; (2) groundwater proximity; and (3) the morphology and organization of passages (Engel 2010). Fig. 1.1 outlines the types of caves and the original forming processes for those caves (Engel 2010). To date the most extensively studied caves for their microbial potential are karstic or limestone caves. Recently, studies have diversified and significant attention has been given to evaluating microbiota in lava tube and basaltic caves as a way to evaluate potential astrobiological environments. These studies were initiated after analysis of known Martian environmental characteristics' resemblance to basaltic caves on Earth. It is hoped that studies into terrain caves will provide insights into the potential microbial composition, diversity, dynamics and relationships, and evolutionary importance of earlier microbial

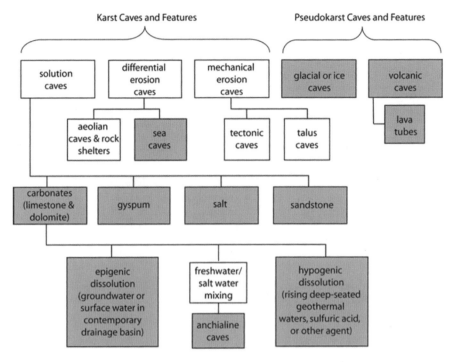

Fig. 1.1 Common types of caves and the processes responsible for forming them. *Gray* boxes indicate which cave types have been studied to characterize the microbiology, predominately using molecular methods (Engel 2010)

life on Earth and life on other planets (Pedersen 2000; Boston et al. 2001; Léveillé and Datta 2009; Storrie-Lombardi and Sattler 2009; Storrie-Lombardi et al. 2009). Additionally, studies in sulfidic and ice caves have also helped elucidate a complex microbial community structure in what is seemingly an inhospitable environment to most life forms (Priscu et al. 1999; Hose et al. 2000; Rohwerder et al. 2003; Borda and Borda 2006; Borda et al. 2004; Chen et al. 2009; Rusu et al. 2011).

Challenges and Methods in Evaluating Cave Microbial Communities

Using Culture Techniques to Evaluate Populations

Due to the lack of contemporary knowledge of mineral–microbe interactions in caves in the early stage of studies, it was not possible to decipher the community composition in those early population studies using only culturing techniques.

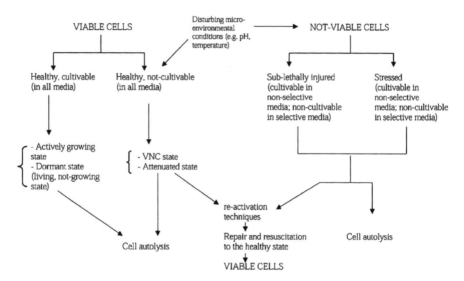

Fig. 1.2 Ecophysiological phenomena related to the cultivation ability of microbial cells in natural ecosystems (Giraffa and Neviani 2001)

The population analysis resulting from these early studies reflected limitations of a media only approach to survey microbial populations. It is probably that only an estimated 1% of cave microbe flora was identified in these studies (Giovannoni et al. 1990; Ward et al. 1990; Amann et al. 1995; Hugenholtz et al. 1998). Fig. 1.2 shows ecophysiological phenomenon related to cultivability of microorganisms in natural ecosystems (Giraffa and Neviani 2001).

As previously stated, it is evident that not all microorganisms possess the same level of readiness in cultivability (ability to be cultivated on an artificial medium in vitro). In an ecosystem there are two types of cells: viable cells (culturable) and nonviable (non-culturable) cells. Typically, when adverse conditions such as low pH, low temperature, limited nutrients, and other stresses are present, viable cells make a shift to their non-cultivable state. In this state they remain viable, but non-cultivable (VNC) (Giraffa and Neviani 2001; Vartoukian et al. 2010). There are other explanations for the inability to culture microorganisms in these situations including the following: (1) they are of low prevalence and/or are slow-growing; (2) they have a resistance to being cultured in isolation as a monoculture (i.e., they need some signaling molecules from other species as in cooperative relationships to be able to grow); and (3) they remain under certain constraints such as dormancy or under a temporary state of low metabolic activity (Vartoukian et al. 2010). These can lead to challenges in evaluating microbial communities through culture techniques, since caves are limited in nutrients, but are rich in chemically complex mineral and metal compositions and limiting in organics; medium developed for these organisms needs to be specially formulated to reflect the needs of cave microflora. Likewise, growth conditions must be developed and optimized for slow-growing organisms. There are successful studies undertaken using low-nutrient media to isolate bacteria from diverse environments including caves and acidic Sphagnum peat (Personal communication with Barton H 2009; Pankratov et al. 2008). These limitations may make culturing cryptic indigenous flora impractical as a way to comprehensively evaluate microbial communities, but may be used to evaluate targeted populations within the study. Culture techniques can be thought of as a way to obtain general information of specific metabolic and biochemical properties of the community to aid in further identifying and characterizing certain targeted groups of microorganisms in any environments. This discrepancy of the microbial culturability in environmental samples can underestimate the whole context when studying the composition of an entire microbial community. Furthermore, because of the complex nature of the environment, the difficulties growing a representative community of organism, the tendency to select for faster growing species, and the difficult growth characteristics for the vast majority of the microflora, the results from any culture method are inconclusive. Even in relatively limited studies with growth-enhanced medium it has been proven difficult to truly obtain pure culture from some volcanic cave samples even after the initial isolation on complex medium (Cheeptham, unpublished results; Rule et al. 2011). That is not to say such investigation fails to provide useful community information. The data is relevant and is useful, provided the results report the reflected limitations of the study.

In any given microbial community, caves included, there are two main types of microorganisms that can be cultivated in vitro: fast- and slow-growing organisms. At a given isolation and incubation time period, slow-growing microorganisms often are overlooked in cultural analyses (Vartoukian et al. 2010). Even when physical and nutritional requirements are understood, it is often difficult to evaluate a community because of the tendency to easily isolate fast-growing species over slow-growing ones. This could be due to the lack of knowledge of the metabolisms of slow-growing microorganisms (Personal communication with Stemke D 2011). This bias against isolation of slow-growing microorganisms through culturing method has been documented (Engel et al. 2001).

To fully characterize the physiology of any given species of microorganism, it is generally necessary to obtain the organism in pure culture (Vartoukian et al. 2010). This thought is changing; we are entering an era of advanced molecular techniques that are being developed to identify the physiology of constituent species in environments. The pure culture technique is useful for studies on metabolic requirement and viability of some constituent microorganisms. However, usage of this culturing technique alone for community study limits utility (Engel et al. 2001). Therefore a polyphasic analysis should be employed that uses an integrated approach that combines culture-dependent and culture-independent methods to best analyze the composition and diversity of a given microbial community (Smalla 2004). Such an integrated approach will provide a more complete understanding of the microbial community's structure and diversity.

Using Culture-Independent Methods to Evaluate Populations

In the last 15 years, molecular based techniques have been reported to provide a more comprehensive representation of bacterial communities than culture techniques in soil environments (Pace 1997; Dahllöf 2002; Rastogi and Sani 2011). Focusing on cave environments, there are many factors needed to be considered when evaluating microbial communities in these chemically and physically unique environments. Cave environments possess heterogeneity of geological characters, including those of speleothems (stalactites, stalagmites, moonmilk, soda straw), walls, and rock samples which further add to the challenges presented for evaluating the community structure in caves (Northup and Welbourn 1997: Macalady and Banfield 2003). Given the heterogeneity of geological sample collections from caves, when culture-independent methods are the choice of the study, considerations in the sampling process and the methods of DNA extraction procedures are critical (Moser et al. 2003; Bent and Forney 2008; Hirsch et al. 2010). Methods in DNA extraction and evaluation through enhanced microbial community analysis have been widely reported (Gurtner et al. 2000; Northup et al. 2003; Barton et al 2006; Maukonen et al. 2012). These methods provide basic sequence analysis to evaluate soil microbial communities from different environments. If a molecular phylogenetic approach using DNA extraction, PCR, 16S rRNA sequencing, and/or other techniques is used for a microbial community composition analysis, a different set of constraints from culturing issues need to be addressed. DNA from these samples is typically in low concentrations and these samples generally have high concentrations of natural DNA-binding agents. Both of these constraints have been addressed in cave communities studies carried out by Barton and colleagues (2006) and Herrera and Cockell (2007). Barton developed a protocol for successfully extracting DNA from low-biomass carbonate rock collected from Jack Bradley Cave in Kentucky and Carlsbad Cavern (New Mexico, USA). This method improved upon other isolation protocols by more effectively extracting DNA extraction from caves that are calcium-rich geologic samples through the use of poly-dIdC, a synthetic DNA molecule. Their improved DNA extraction protocol involved crushing rock sample in a sterile pestle and mortar, and then extraction with stabilization buffers, lysozyme to lyse cells, and poly-dIdC to bind free DNA. Incubation and adding of Proteinase K and sodium dodecyl sulfate (SDS) were done for increasing the purity of the extracted DNA. Further steps were done and an additional poly-dIdC step was included to separate more DNA molecules from calcium, a natural DNA-binding cation abundant in many cave samples. The calcium was then removed through a Slide-A-Lyzer® MINI dialysis unit. This procedure effectively eliminated calcium inhibitors that interfere with downstream PCR amplification. By removal of cations from the samples it was possible to recover 13–14 ng of DNA/μl. The recovered amount of DNA by this improved protocol is from the total number of bacterial cells from these samples which were found to be 10^5–10^6 cells/g of samples. This protocol was successfully used to recover high-quality DNA from samples that were rich in calcium, manganese, or silicates (Barton et al. 2006).

The use of the synthetic DNA molecule, poly-dIdC, was very innovative and useful as an example of such studies. However, it is doubtful that this method can be universally applied to all types of cave samples. Each cave sample is unique in its own physical and chemical characteristics and may require specific modification of purification methods to handle local in situ compounds to best isolate DNA for community analysis. The need in looking at various studies done to find an appropriate DNA extraction protocol to isolate DNA from various volcanic soils is reviewed by Herrera and Cockell (2007). Table 1.1 demonstrates some examples of appropriate DNA extraction techniques used for geological samples from various cave sites. From the table, in Brown and Wolfe's work, a comparison evaluating purification of DNA between CTAB with chloroform purification, UltraClean® (MoBio Laboratories) DNA extraction kit, and FastaDNA® extraction kit (MP Biomedicals) was undertaken. They reported recovered DNA concentrations ranging from 41 to 72 μg/ml from acidic hydrothermal volcanic samples (pH from 1 to 5.8) and 43.2 μg DNA/g sediment recovered through a DNA–RNA extraction method (Purdy et al. 1996).

Unpublished work in our laboratory evaluated three commercially available DNA extraction kits, namely, UltraClean®, PowerSoil®, and PowerLyzer™ (MoBio Laboratories), for their ability to recover DNA from a volcanic floor rock sample (sample#5.2, HelMcKen Falls cave in Wells Gray Provincial Park, British Columbia, Canada). In this study it was found that PowerLyzer™ Powersoil DNA Isolation Kit yielded the highest concentration of DNA (0.30 μg/mL) from the sample while the UltraClean® and PowerSoil® recovered 0.24 μg/mL and 0.165 μg/mL, respectively. These samples are rich in Ca, Fe, Mg, Na, and K (Rule et al. 2011). The low amount of DNA recovered could be caused by the high amount of natural DNA-binding agents (cations observed in the rock sample) and DNA sequestration in hard mineral matrix which are often limits found in recovery of DNA from volcanic samples (Herrera and Cockell 2007). A clear message from the reviews and the independent studies suggests the importance of understanding unique characteristics of cave samples' chemistry prior to attempting to isolate DNA from the sample.

Currently Available Techniques

As stated earlier, there are a wide number of methods employed to detect, identify, and characterize microorganisms in microbial community assessments. These methods can be classified into two main categories: (1) culture-based methods and (2) culture-independent methods. Culturing methods are of those that use either selective or enrichment isolation media and growth conditions to culture microorganisms from environmental samples. However, this includes limitations as previously described. Microbial community composition analysis using a cultivation-independent approach typically involves extraction of marker compounds that are useful in representing the community (Leckie 2005). The useful markers should also be relatively stable and should be variable across interested communities of organisms.

Table 1.1 Examples of DNA extraction method development from a large variety of volcanic environments during the last decade (Herrera and Cockell 2007)

Environmental sample	Chemical characteristics	Bacterial level (cells/g)	DNA extraction method	Extracted DNA yield and quality	Reference
Andisol volcanic ash soil	Clay mineral, pH between 4.84 and 6.20, organic	10^9	*Two lysis procedures*: (bead-beating + 10 min heating 60°C) and (bead-beating + extraction buffer phenol–chloroform–isoamyl), and adding skim milk to the bead-beating to minimize DNA degradation and absorption to soil particles *DNA extraction method*: FastDNA Spin kit for soil (Q-BIOgene)	Detectable DNA for all samples	Takada Hoshino and Matsumoto (2005)
Low-biomass carbonate rocks	Rich in calcium, manganese, or silicates (DNA-binding cations)	10^5–10^6	*Crushing*: Flame sterilized Plattner's pestle and mortar *Extraction Buffer*: EGTA (strong calcium chelator) + lysozyme + poly-dIdC blocking agents of $CaCO_3$ reactive surfaces *Purification*: Phenol–Chloroform with poly-dIdC adding before precipitation and dialyze against EGTA (to remove calcium inhibitors of PCR, instead of gel purification which can be a contamination source)	13–14 ng DNA/µl	Barton et al. (2006)

Sample	pH	Method	Results	Reference
Modern calcareous microbialites	Not given	*Disruption matrix methods:* –Pulverization using a sterile mortar and pestle (but destruction of spatial architecture of microbial communities) –Acid dissolution using sterile HCl solutions with increasing acidity from 5.0 to 0.8 –Chelator-mediated dissolution with sterile ethylene diaminetetraacetic acid (EDTA disodium salt dehydrate) solutions increasing concentrations from 500 to 675 mM at pH 5 *DNA extraction method:* UltraClean Plant DNA Isolation Kit (MoBio Laboratories)	–Pulverization: 3.5 µg DNA/g –Acid dissolution: No detectable DNA –EDTA dissolution: 7.8 µg DNA/g	Wade and Garcia-Pichel (2003)
Acidic hydrother-mally modified volcanic soils	pH between 1.4 and 3.3	*Comparison between several methods:* –FastPrep (Qbiogene) –FastPrep PEG/lysosyme modification –PowerSoil DNA isolation Kit (MoBio Laboratories) –Chemical cell lysis (Xanthogenate–SDS buffer) –Chemical-Enzymatic cell lysis (Phenol) –Mechanical-Chemical-Enzymatic cell lysis (phosphate, SDS, Chloroform, bead beater)	None of the six extraction methods resulted in visible DNA on gel electrophoresis	Henneberger et al. (2006)

(continued)

Table 1.1 (continued)

Environmental sample	Chemical characteristics	Bacterial level (cells/g)	DNA extraction method	Extracted DNA yield and quality	Reference
Acidic hydrothermal volcanic environments	pH between 1 and 5.8	Not given	*Comparison between several methods:* –CTAB with Chloroform purification –UltraClean (MoBio Laboratories) –FastaDNA (MP Biomedicals)	CTAB=FastaDNA>UltraClean, except for three sites (with DNA amounts between 41 and 72 µg/ml)	Brown and Wolfe (2006)
Acidic volcanic rocks	pH 1	Not given	UltraClean DNA extraction from crushed rocks (Mobio Laboratories)	3.1 mg DNA/g crushed rock	Walker et al. (2005)
Saline Mud Volcano	pH 6.5, NaCl dominant brines with 98.0‰. Rich in Al, Fe, Pb	1.8×10^5	FastDNA Spin kit for soil (Q-BIOgene)	Not given	Yakimov et al. (2002)

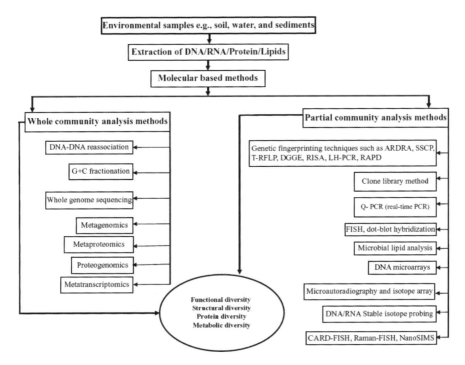

Fig. 1.3 A comprehensive culture-independent molecular toolbox to characterize the structure and functional diversity of microorganisms in the environment (Rastogi and Sani 2011)

The markers should also contain phylogenetic information to be used to make profiling less ambiguous. One example of these biomarker compounds are targeted DNA or RNA molecules. This technique begins with nucleic acid (DNA/RNA) extraction, amplification using PCR (RT-PCR, LH-PCR, c-PCR, AP-PCR), microbial community profiling by means of Genetic Fingerprinting Techniques (ARDRA, RISA, DGGE/TGGE, T-RFLP, or SSCP) and/or sequencing, and molecular phylogenetic analysis. Fundamental details of each method will not be discussed here since they have already been reviewed elsewhere (Tiedje et al. 1999; Rondon et al. 1999; Hill et al. 2000; Giraffa and Neviani 2001; Kirk et al. 2004; Dahllöf 2002; Smalla 2004; Leckie 2005; Malik et al. 2008; Hirsch et al. 2010; Rastogi and Sani 2011).

With culture-independent or molecular based methods, once the extraction of DNA, RNA, proteins, or lipids is completed, there are several specific tools available to characterize structure and function of microorganisms in a given environment. Fig. 1.3 presents a very useful view of current culture-independent molecular tools. The figure shows that there are two main streams of molecular based methods depending on capacity of unlocking structure and function of microbial community diversity: (1) whole community analysis methods and (2) partial community analysis methods (Rastogi and Sani 2011). Additionally, Table 1.2 lists some of the

Table 1.2 Summary of the most widely used culture-independent techniques and their applications in microbial ecology (modified from Giraffa and Neviani 2001)

	Taxonomic resolution	Applications to microbial ecology
Molecular based techniques		
(a) Genetic fingerprinting of microbial communities		
DGGE/TGGE[a]	Community members (genus/species level)	Dynamics between microbial populations in different natural environments
SSCP[b]	Community members (genus/species level)	Mutation analysis; dynamics between microbial populations in different natural environments
Other PCR-based methods		
T-RFLP[c]	Community and population members (genus, species, strain level)	Strain identification; dynamics between and within microbial populations in soils, activated sludge, aquifer sand, termite gut
LH-PCR[d]	Community members (genus/species level)	Dynamics between microbial populations in aquatic and soil microbial environments
PCR-ARDRA[e]	Community members (species level)	Automated assessment of microbial diversity within communities of isolated microorganisms
RISA/ARISA-PCR[f]	Community members (species level)	Estimation of microbial diversity and community composition in freshwater environments
AP-PCR[g]	Population members (strain level)	Automated estimation of microbial diversity (typing) within lactic acid bacteria populations
AFLP[h]	Community and population members (genus, species, and strain level)	Automated estimation of microbial diversity within communities (species composition) and populations (typing) of various Gram-positive and Gram-negative bacteria
(b) Competitive PCR	Community members (species level)	Detection of microbial cells into the VNC state in freshwater samples

Nonmolecular-phylogenetic-based techniques

Technique	Taxonomic level	Description
(c) Fluorescence in situ techniques		
Fluorescence in situ hybridization (FISH)	Community members (species level)	Detection of viable cells within bacterial communities from environmental samples or food ecosystems
Fluorescence in situ PCR	Community members (species level)	Detection of viable, slow-growing cells within bacterial communities, particularly pathogens in clinical specimens
(d) Electron microscopy		
Scanning and transmission electron microscopes (SEM and TEM)	N/A (not by itself)	As an aid to further analysis for additional positive identification. Coupled with labelling techniques can be very useful
(e) PLFA profile		
Fatty acid methyl ester analysis (FAME)	Community members (genus and species level)	Identification of pathogenic bacteria in food samples and clinical specimens and microorganisms in different ecological samples. Used in microbial source tracking. If used with stable isotope probing, dynamics of soil microbes can be tracked. No culturing is needed
(f) %GC content	N/A (not by itself)	Global view of community diversity
(g) Stable isotope probing (SIP)	N/A	Linking microbial phylogeny with function. Lack sensitivity and should not be used by itself. Can provide enrichment biases
(h) Microarray technologies	Community members (genus)	Identify microorganisms and determine ecological role. A number of weaknesses such as lack of specificity, sensitivity, and quantification

[a] Denaturing gradient gel electrophoresis thermal gradient gel electrophoresis
[b] Single-strand conformation polymorphism
[c] Terminal-restriction fragment length polymorphism
[d] Length heterogeneity-polymerase chain reaction
[e] Polymerase chain reaction-amplified ribosomal DNA restriction analysis
[f] Ribosomal spacer analysis/automated ribosomal spacer analysis-polymerase chain reaction
[g] Arbitrarily primed-polymerase chain reaction
[h] Adaptor fragment length polymorphism

techniques and applications in more details. Each of these techniques comes with advantages and limitations. With the advent of improved primers, 16S rRNA sequencing has become widely the method of choice for evaluating bacterial communities in environmental samples. Many studies incorporate the DGGE approach since it allows the separation of partial 16S rDNA amplicons according to their sequence and the simultaneous fingerprinting of bacterial communities of many different samples (Muyzer et al 1993). Other biomarkers such as proteins, phospholipid fatty acids (PLFA), and other taxa-specific cellular constituents have been used to assess microbial community structure and functions (Kirk et al. 2004; Leckie 2005; Hirsch et al. 2010). One important issue regarding the use of these biomarkers is that these molecules can be directly extracted from the environmental samples negating the need for using culturing techniques to acquire a pure culture of microorganism for microbial diversity studies. For instance, phospholipids and fatty acids are important components found in living cell membranes and the lipid composition in the membranes can be changed according to the environmental conditions. As a result, patterns and changes (as in profiles) of phospholipid-derived fatty acids are found to be a useful tool to understand community structure and physiological state of specific microbial groups (Hill et al. 2000; Duran et al. 2006; Lerch et al. 2009; Slabbinck et al. 2009; Malik et al. 2008).

As outlined in Table 1.2, there are a number of culture-independent approaches which are not related to culture-independent molecular phylogenetic studies (nucleic acid-based identification). These include transmission and scanning electron microscopy, Fluorescence in situ Hybridization (FISH), microbial lipid analysis, determination of C-G content, microautoradiography, isotope array analysis, DNA microarray analysis, and DNA/RNA stable isotope probing. Details of each technique are reviewed elsewhere (Giraffa and Neviani 2001; Rastogi and Sani 2011).

Additional innovative methods have been developed to solve the technical difficulties associated with deducing cave microbial community structure. As previously discussed, diverse sampling environments require diverse nucleic acid recovery systems. DNA extraction and recovery are influenced by chemical, physical, and biological factors of the samples. Chemical factors that limit DNA extraction efficiency particularly in volcanic samples include (1) binding of DNA with naturally presented sample components such as cations; (2) sequestration of DNA in hard mineral matrix; and (3) degradation of DNA in acidic conditions. With volcanic environments, the problematic DNA binding lies with Andisol, or volcanic ash soils. This clay mineral interferes with DNA recovery. Difficulties in DNA extraction and recovery were observed when first using a traditional DNA extraction kit (Takada Hoshino and Matsumoto 2005). The traditional DNA extraction kit used in these assays involved direct cell lysis. In this approach vigorous glass-bead shaking of a 500 mg of a soil sample in buffer containing detergents prior to DNA recovery and purification was the method used. Further isolations were successfully done by adding skimmed milk (found to be most effective) or RNA as an adsorption competitor (Takada Hoshino and Matsumoto 2006; Volossiouk et al. 1995 and Frostegård et al. 2011).

DNA extraction involves microbial cell extraction which is undertaken through cell separation, cell lysis, and DNA purification. A number of studies featuring comparison between procedures such as pulverization, acid dissolution, and disruption of matrices and liberating cells from mainly volcanic samples have been evaluated (Herrera and Cockell 2007). Extractions from rocky samples require the removal of nucleic acid from stone, which must be first pulverized to access nucleic acids from the matrix. Once pulverized, the issues of sequestration of the nucleic acids from cations are still pertinent and a protocol must be used to further isolate the released DNA. With acid dissolution and other cell-liberating techniques, it is important to recognize further potential for DNA degradation in the sample which will lower DNA extraction ability.

There are generally two ways to extract DNA from these isolates: (1) through an indirect purification whereby cells are firstly separated from soil or rock samples and then cells are lysed for DNA purification and (2) a direct purification protocol whereby cell lysis is done directly in the soil or rock samples before DNA extraction occurs. The first approach is found to be generally cumbersome in both time and effort with surprisingly low yield. In addition, it was found that some specific groups of microorganisms, such as ammonia-oxidizing bacteria, can be especially difficult to remove directly from soil particles (Herrera and Cockell 2007). This trapping of cells within the matrix can lead to biases in the DNA extraction method and the results gained can be very misleading. The second approach is more widely used because it generally results in higher DNA yields.

Recently, in situ cell disruption methods including both physical and chemical protocols which include a subset of grinding–freezing–thawing, sonication, bead-beating, heating treatment, and/or detergent have been adapted for nucleic acid purification (Robe et al. 2003). Studies comparing procedures for efficient DNA extraction from different geological samples showed different levels of efficiency and quality of DNA recovered. In summary, it was demonstrated that depending on unique characteristics of each sample, different DNA extraction procedures and steps involved in such endeavors needed to be tested to evaluate which procedure was the most appropriate to maximize extract purification (Herrera and Cockell 2007). After DNA extraction is done, PCR is used to amplify microbial DNA. Recovered DNA is then used for DGGE or 16S rRNA sequencing or further evaluated through one of the other appropriate protocols for the study. Not only DNA extraction protocols but also the length of storage of DNA extraction kits may contribute to many biases in interpretation of composition and diversity of a microbial population (Maukonen et al. 2012).

In addition to chemical interference influencing DNA extraction in volcanic sample, there are also physical limitations. In volcanic environments, rock samples often include high content of glass/silica. Therefore, traditional glass bead treatment to disrupt sample particles for cell liberation is ineffective because both DNA shearing can occur and there are mechanical difficulties in the isolation of silica (Herrera and Cockell 2007). In these samples, other approaches such as physically pulverizing the rock and further grinding in a sterile mortar are often used as initial preparation steps (Thorseth et al. 2001; Lysnes et al. 2004). These crude preparation steps potentially lead to increased risk of contamination. Since both DNA quantity and quality are vital to determine the community composition, it is important to pair the best protocol with the chemical properties of the sample.

Besides the challenges of DNA extraction, biases and preferences that occur in PCR amplification is another issue that needs to be considered when evaluating microbial communities (Kirk et al. 2004). Underestimation of microbial diversity in a given sample by PCR- and RFLP-based methods can be caused by insufficient cell lysis and the preferential and selective amplification of 16S rRNA gene fragments (Zhou et al. 2007). With insufficient cell lysis, biases inevitably result from either chemical or physical lytic treatments of cells for DNA extraction. For example, preferences of DNA isolation are reported between differences of Gram-negative and -positive and archeal cell wall components. It is known that Gram-negative bacteria are prone to lyse more easily than Gram-positive bacteria in lysozyme extraction protocols (Takada Hoshino and Matsumoto 2005).

It is also important to consider the predisposition of certain steps in sample prep-aration to influence DNA recovery. Differences in isolation protocols involved in DNA recovery and extraction of bacterial, archeal, as well as eukaryotic microbiota are critical in the assessment of which molecular tools can most effectively be used to study microbe populations. It is important to understand that although molecular techniques are powerful tools to uncover microbial composition and diversity of a given community these methods have their limitations (Zhou et al. 2007; Giraffa and Neviani 2001; Rastogi and Sani 2011). Because of the nature of DNA amplification through PCR there is going to be a predilection to amplify the more common species of nucleic acids and a retardation of the amplification of rarer spe-cies. Only bacterial populations that make up more than 1% of the total community can be detected by PCR (Muyzer et al. 1993; Murray et al. 1996; Muyzer and Smalla 1998). As such, PCR amplification protocols generally are not going to give a com-pletely accurate representation of the microbial community structure. It should also be understood, therefore, that nucleic acid-based methods alone cannot allow the investigator to make a comprehensive representation of a microbial community. However, PCR can provide a reasonable portrayal of the indigenous flora.

The final challenge in purification of DNA from volcanic samples is the presence of biological contamination. Microorganisms are everywhere. It is difficult to deter-mine how one can be certain that DNA extracted from a complex sample is not contaminated during sample collection, transportation, and/or DNA extraction experimental procedures. Contamination of the sample will compromise the analy-sis of the microbial communities. As Barton and colleagues reported, biological contaminations are of special concern when interpreting data in low-biomass sam-ples from caves (2006).

Biological contamination can occur at any step leading to analysis of the sample. In order to minimize contamination clean-lab procedures are necessary to reduce the risk of potential contamination. Such procedures routinely include bleaching/autoclaving of equipment, instruction of personnel to limit worker contamination, use of a laminar flow hood, and restricted airflow with positive air hoods or glove boxes, these procedures are all precautionary steps necessary to minimize sample contamination. "Clean" laboratories should be physically separated from other microbial/molecular laboratories with separate ventilation systems and nightly UV irradiation is recommended to minimize sample contamination (Willerslev et al. 2004). Biological contaminations can also originate from DNA extraction reagents and commercial kits, even from PCR amplification procedure itself (Barton et al. 2006). In order to establish that a sample is contamination free, it is necessary to have experimental controls (positive and negative) included in every step during sample analysis. Positive controls are often designed with standard microorganisms to easily identify the presence of contaminating organisms while negative controls do not typically have any microorganism to make sure that reagents and chemicals used are clean without contamination.

Microbes Discovered in Caves

Returning once again to the two most commonly studied types of caves, namely, limestone (calcium carbonate) and lava tube (basalts), it should be of little surprise that these types of caves are composed of radically different distinct mineral compositions and consequentially possess different composition and diversity of microbial communities. In essence, the question of "what elements are the major contributors to the diversity of microbe cave life" still needs to be answered. Since to date most comprehensive community analysis has been done in limestone caves there is yet little antidotal evidence that the microbial diversity is significantly different between the two major cave types. Table 1.3 summarizes a number of selected studies in regard to techniques used, and microorganisms found, in karstic and lava tube caves and related subsurface environments. As is evident from these findings, prior to 1995 studies of cave community structure were primarily evaluated via culture techniques. With the development of molecular techniques there was a rapid shift to using molecular techniques as the standard way to conduct community analysis. Trends of studies can also be deduced in Table 1.3, which show a tendency of using a combination of techniques in many studies starting from 2000. Part of the shift to combine culture and non-culture techniques was the result of questions being asked that were best answered through combined analysis (Northup et al. 2011).

Summarizing results of microorganisms found in both karstic and lava tube caves (Table 1.3), the general findings were that certain dominant species trended across the cave systems (notably members of *Actinobacteria, Proteobacteria, Planctomyces, Chloroflexi, Acidobacteria)* while the trend of specialized microorganisms (i.e., *Tiothrix,* flexibacter–Cytophaga–Bacteroides phylum) can be found more in specific geological environments. A comparison of microbial populations in dripping waters and ceiling rocks in karstic caves was undertaken by Laiz and colleagues (1999) who investigated the different microflora of dripping waters and ceiling rocks in Altamira cave (Santillana del Mar, Spain) using Petrifilm aerobic count plates and various selected agar media. Gram-negative rods and cocci

Table 1.3 Review of microorganisms found and methods used in cave and related subsurface microbial community study

	Sample types and environments	Objectives	Techniques	Microorganisms found	Reference
1	Surfaces of the rock art paintings of Atlanterra shelter, Spain	To study bacteria present in the rock art paintings	Hygicult (dipslides) for isolation and total cellular fatty acid methyl esters (FAME) for identification	Members of *Bacillus* species, predominantly *B. megaterium*	Gonzalez et al 1999
2	Surfaces of cave walls from two karstic caves, Altamira and Tito Bustillo in northern Spain	To isolate and identify microorganisms contributing to deterioration of rock arts in caves	Cultural methods (with contact plates, cotton swabs, and porous plastic plugs and various selected isolation media for isolation, while morphological, physiological, and chemotaxonomic characteristics were examined for identification	Members of *Streptomyces* (predominant), *Nocardia, Rhodococcus, Nocardioides, Amycolatopsis, Saccharothrix, Brevibacterium, Microbacterium,* and coccoid actinomycetes (family *Micrococcaceae*)	Groth et al 1999
3	Dripping waters and ceiling rocks in Altamira cave Santillana del Mar, Spain	To study and compare microbial populations in dripping waters and ceiling rocks in a karstic cave (to better understand microbial involvement on rock art deterioration in caves)	Culture method (Petrifilm aerobic count plates and many selected agar media for isolation)	In dripping waters, Gram-negative rods and cocci (*Enterobactericeae* and *Vibrionaceae*) were found while mainly *Streptomyces* were observed in ceiling rock samples	Laiz et al 1999
4	Various cave samples (stream sediments, black mud, water drips, snottites (stalactitic forms)) from Cueva de Villa Luz, Tabasco, Mexico	To study microbial contribution in the development of acidic environments and cave enlargement in Cueva de Villa Luz (hydrogen sulfide-rich karst environment)	Combination of methods; culture-dependent and -independent approaches (various isolation media, SEM, ESD, 16S rRNA sequencing)	A number of sulfur-oxidizing and -reducing bacteria were found depending on sample types. In summary, *Thiobacilli* spp., *Thiothrix* spp., and *Acidimicrobium ferrooxidans*	Hose et al 2000

#	Sample	Objective	Methods	Findings	Reference
5	Surface materials from wall paintings of Catherine Chapel, Castle of Herberstien, in Austria and of St. Martin's church, Greene, in Germany	To investigate and compare bacterial diversity in two medieval biodeteriorated wall paintings	Combination of methods; molecular and culturing techniques (DGGE, PCR, 16S rRNA sequencing, selected isolation media for heterotrophic bacteria, and FAME)	A number of members of Actinobacteria and Proteobacteria (from molecular techniques) while Bacillus, Paenibacillus, Micrococcus, Staphylococcus, Methylobacterium, and Halomonas were seen (from culturing techniques). NOTE: No similar organisms could be detected by the cultivation and the molecular approaches	Gurtner et al 2000
6	Water samples, microbial mats, and biofilms from pools and spring of the Cesspool cave, Virginia, USA	To determine acid-producing bacterial community composition and its geomicrobiological functions in cave	Combination of methods; molecular and liquid enrichment culturing techniques (16S rRNA sequencing, SEM, and sulfur-oxidizing bacteria enrichment liquid culture for 3 months)	Genus *Thiothrix* and the flexibacter–Cytophaga–Bacteriodes phylum and maybe *Helicobacter* or *Thiovulum* group for some remaining strains	Engel et al 2001
7	Microbial mentles and water from Weebubbie cave, Nullarbor Plain western Australia	To examine a cave biomineral of novel morphology	FESEM, TEM, XRD, FTIR, and SAED	Bacterial filaments associated with calcite crystals	Contos et al 2001
8	Pull fingers from Hidden cave, New Mexico, USA	To preliminarily observe for possible bacterial involvement in spele-othem formation	SEM, X-ray diffraction, and stable isotopic analyses	Calcified filaments and micro-rods observed in samples indicating filamentous bacteria according to the size and morphology	Melim et al 2001
9	Speleothems (active stalactites, wall concre-tions), rock walls, ceiling, and soils from galleries of Grotta dei Cervi, Porto Badisco, Italy	To investigate culturable heterotrophic microbial communities present in the cave	Tryptone soy agar, malt-yeast extract-agar, starch-casein agar, glycerol-asparagin-agar, and water-agar (incubation up to 8 weeks)	Members of *Agromyces*, *Arthrobacter*, *Rhodococcus*, *Streptomyces*	Groth et al 2001

(continued)

Table 1.3 (continued)

	Sample types and environments	Objectives	Techniques	Microorganisms found	Reference
10	Speleothem samples in basaltic sea caves, Kauai, Hawaii, USA	To describe kerolite and associated clay minerals	X-ray diffraction, energy dispersive X-ray microanalysis, SEM, TEM	Bacteria and cyanobacteria	Léveillé et al. 2002
11	Bacterial mats (slimes) in Four Windows cave, a lava tube in El Malpais National Monument, New Mexico, USA	To characterize microbial communities of the lava wall slime using culture-independent methods	16S rRNA sequencing (with RFLP) and SEM	Members of *Actinobacteria, Chloroflexi, Verrucomicrobia,* and *Betaproteobacteria*	Northup et al 2004
12	Bat guano samples in various Japanese caves (20 limestone and volcanic caves)	To examine yeasts in bat-inhabited caves	YM agar with selected antibiotics for isolation and D1/D2 26S rRNA sequencing for identification	Members of *Trichosporon* species, *Candida palmioleophila, C. lusitaniae, Debaryomyces hansenii, Hanseniaspora* spp., *Saccharomyces cerevisiae, S. kluyveri, Williopsis californica, Zysosaccharomyces florestinus,* and *Crytococcus podzolicus*	Sugita et al 2005
13	Various geologic samples from Jack Bradley Cave, Kentucky, and Carlsbad Cavern, New Mexico, USA	To describe an improved protocol to recover extremely low amounts of DNA from calcium-rich geologic samples	Improved DNA extraction protocol (use of Poly-dIdC, EGTA, and Slide-A-Lyzer MINI small-volume dialysis cassettes)	Contamination-free, genomic DNA from a low-biomass, $CaCO_3$-rich cave samples	Barton et al 2006
14	Rock varnish and powders from Whipple Mountains of Sonoran Desert, Arizona, USA	To characterize microbial rock varnish communities in a semiarid to arid region	Molecular approach (DNA extraction, PCR, 16S and 18S rRNA, Cloning, RFLP, and phylogenetic analysis), SEM, and PLFA	Mostly *Proteobacteria* with *Actinobacteria,* eukaryota and a few members of *Archaea*	Kuhlman et al 2006

15	Soil samples from Niu cave, an earth-cave, in Guizhou Province, China (it is in karst area but the cave is unique in composition of soil rather than limestone or calcareous rocks)	To investigate the bacterial diversity in Niu cave	Molecular approach (DNA extraction, PCR, 16S rRNA, Cloning, and phylogenetic analysis)	Members of *Proteobacteria, Acidobacteria, Planctomyces, Chloroflexi* (Green non-sulfur bacteria), *Bacteroidetes, Gemmatimonadetes, Nitrospirae,* and *Actinobacteria*	Zhou et al 2007
16	Water and bacterial mats (biofilms) samples from a subsurface thermal spring, Franz-Josef-Quelle, in Bad Gastein, Austria	To determine communities of *Archaea* and *Bacteria* in a subsurface radioactive thermal spring in Austrian central Alps and to find ammonia-oxidizing *Crenarchaeota*	Molecular approach (DNA extraction, PCR, 16S rRNA, Cloning, RFLP, and phylogenetic analysis) and SEM	*Crenarchaeota* and *Euryarchaeota* Members of the *Proteobacteria* (α, β, γ, and δ), *Bacteroidetes,* and *Planctomycetes*	Weidler et al 2007
17	Limestone, sandstones, and granite cliffs samples in semiarid montane zones of Rocky Mountain in Colorado and Wyoming, USA	To determine compositions of the selected endolithic communities with cultural independent, rRNA-based molecular phylogenetic methods	Molecular approach (DNA extraction, PCR, rRNA, Cloning, RFLP, and phylogenetic analysis)	Abundant and diverse communities were found, most common ones were members of the *Actinobacteria, Cyanobacteria,* and *Proteobacteria*	Walker and Pace 2007
18	Biofabrics in a geothermal mine adit (karst environment), Colorado, USA	To study microbial involvement and influences in mineral deposition	16S rRNA (molecular phylogenetic analyses), PCR (functional gene analysis), ESEM, and EDS	*Thermus* spp., members of Crenarchaeota, *Chloroflexi,* and *Gammaproteobacteria*	Spear et al 2007

(continued)

Table 1.3 (continued)

	Sample types and environments	Objectives	Techniques	Microorganisms found	Reference
	Rock varnish and surrounding soils from Yungay region in Atacama Desert, Chile	To characterize microbial rock varnish communities in a hyperarid region	Molecular approach (DNA extraction, PCR, 16S rRNA, Cloning, RFLP, and phylogenetic analysis) and ATP assay	*Alphaproteobacteria, Betaproteobacteria, Gammaproteobacteria, Bacteroidetes, Chloroflexi* (Green non-sulfur bacteria), *Gemmatimonadetes, Actinobacteria,* and *Cyanobacteria*	Kuhlman et al 2008
19	Cave soil sample, Phanangkoi, Northern Thailand (karstic cave)	To isolate less intensively studied bacteria with antimicrobial activity for potential drug discovery	Humic acid-Vitamin (HV) agar with selected antibiotics for isolation and 16S rRNA with chemotaxonomic and phynotypic characteristics for identification	A new strain of *Spirillospora albida*, an actinomycetal bacterium	Nakaew et al. 2009a, b
20	Cave soil samples, Phanangkoi and Phatup caves, Northern Thailand (karstic cave)	To isolate rare actinomycetes from caves using selective methods	Humic acid-Vitamin (HV), Humic-acid vitamin-gellan gum (HVG), and Starch and Casein (STC) agar with selected antibiotics for isolation and 16S rRNA with chemotaxonomic and phynotypic characteristics for identification	Members of *Spirillospora, Catellatospora, Nonomuraea*	Nakaew et al. 2009a, b

21	Cave popcorn from an unnamed karstic cave in Kentucky, USA	To investigate bacterial calcium carbonate precipitation in cave environments	Conventional isolation on B-4 and B-4C media and 16S rRNA sequencing (identification) and SEM	Members of *Alpha-, Beta-, and Gammaproteobacteria Firmicutes* and *Actinobacteria*	Banks et al 2010
22	Rock walls, speleotherms surfaces and soil samples from Turkish karstic caves, Turkey	To isolate actinomycetes from caves	Starch Casein agar with selected antibiotics for isolation and chemotaxonomic and phynotypic characteristics for preliminary identification	Members of *Streptomyces*	Yücel and Yamaç 2010
23	Speleothems (stalactites and stalagmites) from Buracos cave (lava tube), Terceira island (Azores), Portugal	To investigate microbial contribution to iron speleothems for a better understanding of roles of microbes in speleothem construction	Microscopy (SEM, LT-SEM, SEM-EDS) and molecular approach (16S rRNA sequencing, PCR-DGGE)	From DGGE, mainly *Proteobacteria* and *Actinobacteria* were found while *Gallionella* sp. and *Leptothrix* sp. were observed from SEM but not DGGE	de los Ríos, Bustillo and Ascaso 2011
24	Rock chips covered with either microbial mats or secondary minerals from basaltic lava caves in Azores (Portugal), Hawaii and New Mexico (USA)	To study microbial communities in microbial mats and mineral-like deposits in lava tube caves	Combination of SEM and molecular methods (DNA extraction, PCR, 16S rRNA sequencing and phylogenetic analysis)	Fourteen phyla of bacteria across three locations (*Actinobacteria, Acidobacteria, Proteobacteria* (α, β, γ, ε, and δ), *Bacteroidetes, Chloroflexi, Nitrospirae, Verrucomicrobia, Gemmatimonadetes, Planctomycetes*)	Northup et al 2011

(*Enterobacteriaceae* and *Vibrionaceae*) were found in dripping water while mainly *Streptomyces* were observed in ceiling rock samples. In 2000, Engel and colleagues revealed that the genus *Thiothrix* and members of the phyla Flexibacter–Cytophaga–Bacteroides and possibly *Helicobacter* or *Thiovulum* groups were discovered from a combination of molecular based and liquid enrichment culturing based methods. 16S rRNA sequencing, SEM, and sulfur-oxidizing bacteria enrichment liquid culture (incubation period for 3 months) were all employed in this study to determine acid-producing bacterial community composition and its geomicrobiological functions in cave water samples, microbial mats, and biofilms from pools and spring of the Cesspool cave (Virginia, USA). Members of the *Proteobacteria, Planctomyces, Chloroflexi, Acidobacteria* (Green non-sulfur bacteria), *Bacteroidetes, Gemmatimonadetes, Nitrospirae*, and *Actinobacteria* were observed through molecular approaches (DNA extraction, PCR, 16S rRNA, Cloning, and phylogenetic analysis) to investigate the bacterial diversity in Niu cave (Guizhou Province, China) (Zhou et al. 2007). Soil samples from Niu cave, an earth-cave, were investigated in this study. Notably, this cave is in a karst area, but the cave is unique in the composition of soil rather than limestone or calcareous rocks. In 2011, Northup and colleagues found 14 phyla of bacteria across three lava tube caves (from three different climate zones: Hawaii/tropical, the Azores/temperate, New Mexico/semiarid) studied including *Actinobacteria, Acidobacteria, Proteobacteria* (α, β, γ, ε, and δ), *Bacteroidetes, Chloroflexi, Nitrospirae, Verrucomicrobia, Gemmatimonadetes*, and *Planctomycetes*. These groups of community organisms were identified using a combination of SEM and molecular methods (DNA extraction, PCR, 16S rRNA sequencing, and phylogenetic analysis) in microbial mats and mineral-like deposits in these lava tube caves. In this study, rock chips covered with either microbial mats and/or secondary minerals from basaltic lava caves in the Azores (Portugal) and Hawaii and New Mexico (USA) were investigated. An extensive collection of scanning electron photomicrographs and spectra of microorganisms found in lava tubes caves is published online at Karst and Cave Studies, UNM Research Center at the University of New Mexico Web site and can be found at http://repository.unm.edu/handle/1928/9480. This site is managed by Michael N. Spilde, Manager, Microbe/SEM Facility, Institute of Meteoritics, UNM, 277-5430, mspilde@unm.edu. Some examples of such images are seen in Fig. 1.4. Fig. 1.5 demonstrates some selected SEM images of not-yet-identified microorganisms in popcorn samples found at Helmcken Falls volcanic cave in Wells Gray Provincial Park, British Columbia, Canada. As previously mentioned in more detail, Gurtner and colleagues (2000) reported that no common organisms were found when comparing cultivated and molecular techniques in a series of degraded cave wall paintings. This discrepancy between microbial community findings between cultivation and molecular methods is also seen in other caves' and non-caves' community analysis studies (Northup et al. 2011; Rheims et al. 1996). This reflects the limitations already noted (Gurtner et al. 2000). These findings reemphasize the relevance of using a combination of approaches to evaluate microbial community composition and diversity.

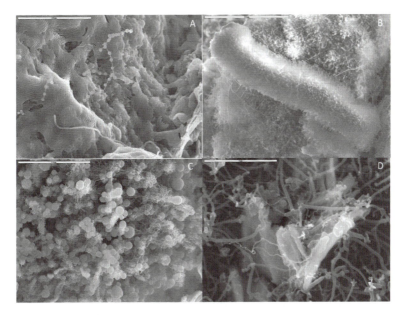

Fig. 1.4 SEM images of microorganisms observed in some speleothems found in lava tubes. (**a**) Beads on a string, filaments, film of ordered rods from white speleothem (Scale Bar: 20 μm, Magnification: 2,300, Etch: No etching, Coating: Au/Pd). (**b**) Large wooly rod shape covered with filaments from gold tan speleothem (Scale Bar: 50 μm, Magnification: 1,100, Etch: No etching, Coating: Au/Pd). (**c**) Spheroids with connecting filaments from gold yellow speleothem (Scale Bar: 10 μm, Magnification: 4,500, Etch: No etching, Coating: Au/Pd). (**d**) Filament tangle with crystal from pink speleothem (Scale Bar: 5 μm, Magnification: 9,000, Etch: No etching, Coating: Au/Pd). Photo copyright: Published with kind permission of Michael Spilde and Diana Northup. All Rights Reserved

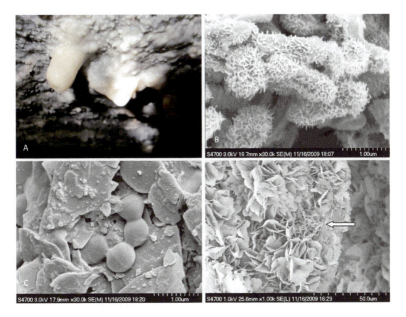

Fig. 1.5 SEM images of not-yet-identified microorganisms observed in some white speleothem (popcorn) found in Helmcken Falls volcanic cave in British Columbia, Canada. (**a**) Cave popcorn. (**b**) Short filaments on top of crystal. (**c**) Fossilized spheroids (connected as a *ring*). (**d**) Long chain of filamentous short rods. Photo Copyright: Naowarat Cheeptham

Summary

In conclusion, caves are poorly understood extreme environments and cave microbiology as a field is still in its infancy. Even with the advent of current techniques, the entire composition and diversity of a microbial community from any environment cannot currently be elucidated. While current technologies are useful to generally evaluate microbial communities in cavern soils, the state of the science is not yet there with any one technique to fully evaluate a given soil or environmental community. A combination of techniques may provide a more realistic representation of the population, presumably until more accurate protocols are developed to assay microbial flora.

Even within these limitations, a contemporary insight into the microbial flora of caves provides an understanding of the formation of a given cave through biomineralization, bioprecipitation, and secondary mineral deposits resulting in speleothems formations. This analysis also instructs geochemists on processes that lead to the degradation of the cave. Geomicrobiological activity interactions in caves were the focus of several studies to identify mineral–microbe interactions involved in this activity (Northup and Lavoie 2001; Engel et al. 2001; Barton et al. 2001; Groth et. al. 2001; Cañaveras et. al. 2001). The value of these studies also informs researchers studying complex interdependence between micro- and macroorganisms to formulate a fundamental understanding of the role of cavern microbe community in the general ecology of the cave.

The need to more fully understand cave microbiology opens possibilities to potential several avenues of research, including exploration of novel biotechnology products, environmental microbiology of caves, geochemistry of caves, and other related disciplines. Fundamental research in this area has led to the discovery of several new genera and species of cave microorganisms such as members of the genera, *Spirillospora*, *Catellatospora*, and *Nonomuraea*, which were recently discovered in a Northern Thailand cave (Nakaew, Pathom-aree and Lumyong 2009a, b). In 2005, *Agromyces subbeticus* sp. nov., isolated from a cave in southern Spain, was reported (Jurado et al. 2005). By the same research group, yet a new species of *Aurantimonas altamirensis* sp. nov, a member of the order *Rhizobiales,* was discovered from Altamira Cave (Jurado et al. 2006). Along with other studies, these newly discovered microbes demonstrated potentials in producing bioactive compounds which have yet to be fully evaluated and potentially exploited (Nakaew et al. 2009a, b; Yücel and Yamaç 2010; Rule et al. 2011; Cheeptham 2011; Sadoway and Cheeptham 2011).

One thing that is evident is that microbiologists need to work in an interdisciplinary capacity to fully elucidate the contributions and complexity of microbial activity. In addition to microbiologists, geochemists, geologists, ecologists, and others, the new field of science offered through the study of metagenomics has the most immediate potential for rapid advancement in the fundamental understanding of the importance of microbial communities in caves. In this field, the term "metagenomics" refers to the study of the genetic material recovered directly from environmental samples (Handelsman 2004). In contrast to traditional microbiology and microbial genome sequencing, metagenomics does not rely upon cultivated clonal cultures to produce a microbial diversity profile in a given natural sample. The potential with this type of analysis is to understand the biological processes undertaken in these systems. As already eluded to, the vast majority of microbial diversity has been missed by both cultivation-based methods and culture-independent molecular based methods. With a metagenomic approach, such as the Sanger sequencing "shotgun" or parallel pyrosequencing protocols, a chiefly unbiased evaluation of all genes from all the members of the sampled communities can be detected. As such metagenomics has recently become a powerful tool to help unlock the hidden world of microorganisms in complex environments. It is hoped that with the development of these tools and a more balanced evaluation of existing screening methods, a more fundamental understanding of the untapped microbiota of these subterranean environments of the Earth will become better illuminated.

Acknowledgements Thanks go to Dr. Douglas Stemke of University of Indianapolis for his invaluable and critical suggestions for the manuscript and to Karen Densky and Jerri-Lynne Cameron of Thompson Rivers University for taking time to proofread.

References

Amann RI, Ludwig W, Schleifer KH (1995) Phylogenetic identification and in situ detection of individual microbial cells without cultivation. Microbiol Rev 59:143–169

Banks ED, Taylor NM, Gulley J et al (2010) Bacterial calcium carbonate precipitation in cave environments: a function of calcium homeostasis. Geomicrobiol J 27(5):444–454

Barton HA (2006) Introduction to cave microbiology: a review for the non-specialist. J Cave Karst Stud 68:43–54

Barton HA, Jurado V (2007) What's up down there? Microbial diversity in caves. Microbe 2:132–138

Barton HA, Luiszer F (2005) Microbial metabolic structure in a sulfidic cave hot spring: potential mechanisms of biospeleogenesis. J Cave Karst Stud 67(1):28–38

Barton HA, Northup DE (2007) Geomicrobiology in cave environments: past, current and future prospectives. J Cave Karst Stud 69:163–178

Barton HA, Spear JR, Pace NR (2001) Microbial life in the underworld: biogenicity in secondary mineral formations. Geomicrobiol J 18(3):359–368

Barton HA, Taylor MR, Pace NR (2004) Molecular phylogenetic analysis of a bacterial community in an oligotrophic cave environment. Geomicrobiol J 21:11–20

Barton HA, Taylor NM, Lubbers BR et al (2006) DNA extraction from low-biomass carbonate rock: an improved method with reduced contamination and the low-biomass contaminant database. J Microbiol Methods 66:21–31

Barton HA, Taylor NM, Kreate M et al (2007) The impact of host rock geochemistry on bacterial community structure in oligotrophic cave environments. Int J Speleol 36:93–104

Bent SJ, Forney LJ (2008) The tragedy of the uncommon: understanding limitations in the analysis of microbial diversity. ISME J 2:689–695

Bonacci O, Pipan T, Culver DC (2009) A framework for karst ecohydrology. Environ Geol 56:891–900

Borda C, Borda D (2006) Airborne microorganisms in show caves from Romania. Biospeleology and physical speleology. Trav Inst Spéol "Émile Racovitza" 43–44:65–74

Borda D, Borda C, Tăma T (2004) Bats, climate, and air microorganisms in a Romanian cave. Mammalia 68(4):337–343

Boston PJ, Spilde MN, Northup DE et al (2001) Cave biosignature suites: microbes, minerals and Mars. Astrobiol J 1(1):25–55

Brown PB, Wolfe GV (2006) Protist genetic diversity in the acidic hydrothermal environments of Lassen Volcanic National Park, USA. J Eukaryot Microbiol 53:420–431

Cañaveras JC, Sanchez-Moral S, Soler V, Saiz-Jimenez C (2001) Microorganisms and microbially induced fabrics in cave walls. Geomicrobiol J 18:223–240

Cardoso P (2012) Diversity and community assembly patterns of epigean vs. troglobiont spiders in the Iberian Peninsula. Int J Speleol 41(1):83–94

Caumartin V (1963) Review of the microbiology of underground environments. Bull Natl Speleol Soc 25(1):1–14 (ISSN 0146-9517)

Cheeptham N (2011) Drugs from the dark. BC caver: The newsletter of the British Columbia Speleological Federation. Winter 2010–2011, 25(1): 25–27

Chen Y, Wu L, Boden R et al (2009) Life without light: microbial diversity and evidence of sulfur- and ammonium-based chemolithotrophy in movile cave. ISME J 3:1093–1104

Contos AK, James JM, Heywood B et al (2001) Morphoanalysis of bacterially precipitated sub-aqueous calcium carbonate from weebubbie cave, Australia. Geomicrobiol J 18(3):331–343

Crecchio C, Gelsomino A, Ambrosoli R et al (2004) Functional and molecular responses of soil microbial communities under differing soil management practices. Soil Biol Biochem 36(11):1873–1883

Dahllöf I (2002) Molecular community analysis of microbial diversity. Curr Opin Biotechnol 13(3):213–217

de los Ríos A, Bustillo MA, Ascaso C (2011) Bioconstructions in ochreous speleothems from lava tubes on Terceira Island (Azores). Sediment Geol 236(1–2):117–128

Dickson G (1979) The importance of cave mud sediments in food preferences, growth and mortal-ity of the troglobitic invertebrates. Natl Speleol Soc Bulletin 37:89–93

Dunbar J, Takala S, Barns SM et al (1999) Levels of bacterial community diversity in four arid soils compared by cultivation and 16S rRNA gene cloning. Appl Environ Microbiol 65(4):1662–1669

Duran M, Haznedaro lu BZ, Zitomer DH (2006) Microbial source tracking using host specific FAME profiles of fecal coliforms. Water Res 40(1):67–74

Engel AS (2010) Microbial diversity of cave ecosystem. In: Barton LL et al (eds) Geomicrobiology: molecular and environmental perspective. Springer Science and Business Media B.V, The Netherlands, pp 219–238

Engel AS, Porter ML, Kinkle BK (2001) Ecological assessment and geological significance of microbial communities from cesspool cave, Virginia. Geomicrobiol J 18(3):259–274

Frostegård Å, Tunlid A, Bååth E (2011) Use and misuse of PLFA measurements in soils. Soil Biol Biochem 43(8):1621–1625

Giovannoni SJ, Britschgi TB, Moyer CL, Field KG (1990) Genetic diversity in Sargasso Sea bac-terioplankton. Nature 345(6270):60–63

Giraffa G, Neviani E (2001) DNA-based, culture-independent strategies for evaluating microbial communities in food-associated ecosystems. Int J Food Micro 67(1–2):19–34

Gonzalez I, Laiz L, Hermosin B et al (1999) Bacteria isolated from rock art paintings: the case of Atlanterra shelter (south Spain). J Microbiol Methods 36(1–2):123–127

Gonzalez JM, Portillo MC, Saiz-Jimenez C (2006) Metabolically active crenarchaeota in altamira cave. Naturwissenschaften 93:42–45

Groth I, Vettermann R, Schuetze B et al (1999) Actinomycetes in karstic caves of northern Spain (altamira and tito bustillo). J Microbiol Methods 36:115–122

Groth I, Schumann P, Laiz L et al (2001) Geomicrobiological study of the grotta dei cervi, Porto Badisco, Italy. Geomicrobiol J 18(3):241–258

Gurtner C, Heyrman J, Piñar G et al (2000) Comparative analyses of the bacterial diversity on two different biodeteriorated wall paintings by DGGE and 16S rDNA sequence analysis. Int Biodeter Biodegr 46(3):229–239

Handelsman J (2004) Soils: metagenomics approach. In: Bull TA (ed) Microbial diversity and bioprospecting. ASM, Washington DC, pp 109–119

Henneberger RM, Walter MR, Anitori RP (2006) Extraction of DNA from acidic, hydrothermally modified volcanic soils. Environ Chem 3:100–104

Herrera A, Cockell CS (2007) Exploring microbial diversity in volcanic environments: a review of methods in DNA extraction. J Microbiol Methods 70(1):1–12

Heyrman J, Mergaert J, Denys R (1999) The use of fatty acid methyl ester analysis (FAME) for the identification of heterotrophic bacteria present on three mural paintings showing severe damage by microorganisms. FEMS Microbiol Lett 181(1):55–62

Hill G, Mitkowski NA, Aldrich-Wolfe L et al (2000) Methods for assessing the composition and diversity of soil microbial communities. Appl Soil Ecol 15(1):25–36

Hirsch PR, Mauchline TH, Clark IM (2010) Culture-independent molecular techniques for soil microbial ecology. Soil Biology and Biochemistry 42(6):878–887

Höeg OA (1946) Cyanophyceae and bacteria in calcareous sediments in the interior of limestone caves in Nord-Rana. Norway Nytt Mag Naturvidensk 85:99–104

Hose LD, Palmer AN, Palmer MV et al (2000) Microbiology and geochemistry in a hydrogen-sulphide-rich karst environment. Chem Geol 169(3–4):399–423

Howarth FG (1983) Ecology of cave arthropods. Annu Rev Entomol 28(1):365–389

Howarth FG (2004) Hawaiian islands, biospeleology. In: Gunn J (ed) Encyclopedia of cave and karst science. Fitzroy Dearborn, New York, pp 417–419

Hugenholtz P, Goebel BM, Pace NR (1998) Impact of culture-independent studies on the emerging phylogenetic view of bacterial diversity. J Bacteriol 180(18):4765–4774

Jones B (2010) Microbes in caves: agents of calcite corrosion and precipitation. Tufas and speleothems: unravelling the microbial and physical controls. University of Alberta, p 9–10.

Jurado V, Groth I, Gonzalez JM et al (2005) Agromyces subbeticus sp. nov., isolated from a cave in southern Spain. Int J Syst Evol Microbiol 55:1897–1901

Jurado V, Gonzalez JM, Laiz L et al (2006) Aurantimonas altamirensis sp. nov., a member of the order Rhizobiales isolated from altamira cave. Int J Syst Evol Microbiol 56:2583–2585

Kirk JL, Beaudette LA, Hart M et al (2004) Methods of studying soil microbial diversity. J Microbiol Methods 58(2):169–188

Koch AL (2001) Oligotrophs versus copiotrophs. Bioessays 23:657–661

Kuhlman KR, Fusco WG, La Duc MT et al (2006) Diversity of microorganisms within rock varnish in the Whipple Mountains, California. Appl Environ Microbiol 72(2):1708–1715

Kuhlman KR, Venkat P, La Duc MT et al (2008) Evidence of a microbial community associated with rock varnish at Yungay, Atacama Desert, Chile. J Geophy Res 113(G04022):14

Laiz L, Groth I, Gonzalez I et al (1999) Microbiological study of the dripping waters in altamira cave (Santillana del mar, Spain). J Microbiol Methods 36:129–138

Lavoie KH, Northup DE, Barton HA (2010) Microbe–mineral interactions. In: Sudhir KJ, Khan AA, Rai MA (eds) Cave microbiology. Science Publishers, Enfield, NH, pp 1–45

Leckie SE (2005) Methods of microbial community profiling and their application to forest soils. For Ecol Manage 220(1–3):88–106

Lerch TZ, Dignac M-F, Nunan N et al (2009) Dynamics of soil microbial populations involved in 2,4-D biodegradation revealed by FAME-based stable isotope probing. Soil Biology and Biochemistry 41(1):77–85

Léveillé RJ, Datta S (2009) Lava tubes and basaltic caves as astrobiological targets on Earth and Mars: a review. Planet Space Sci. doi:10.1016/j.pss.2009.06.004

Léveillé RJ, Longstaffe FJ, Fyfe WP (2002) Kerolite in carbon-rich speleothems and microbial deposits from basaltic caves, Kuai, Hawaii. Clays Clay Miner 50(4):514–524

Lysnes K, Thorseth IH, Steinsbu BO, Øvreås L, Torsvik T, Pedersen RB (2004) Microbial community diversity in seafloor basalt from the Arctic spreading ridges. FEMS Microbiol Ecol 50(3):213–230

Macalady J, Banfield JF (2003) Molecular geomicrobiology: genes and geochemical cycling. Earth Planet Sci Lett 212(1–2):1–17

Malik S, Beer M, Megharaj M et al (2008) The use of molecular techniques to characterize the microbial communities in contaminated soil and water. Environ Int 34(2):265–276

Maukonen J, Saarela M (2009) Microbial communities in industrial environment. Curr Opin Microbiol 12(3):238–243

Maukonen J, Simões C, Saarela M (2012) The currently used commercial DNA-extraction methods give different results of clostridial and actinobacterial populations derived from human fecal samples. FEMS Microbiol Ecol 79(3):697–708

Melim LA, Shinglman KM, Boston PJ et al (2001) Evidence for microbial involvement in pool finger precipitation, hidden cave, New Mexico. Geomicrobiol J 18(3):311–329

Moser DP, Boston PJ, Martin HW (2003) Caves and mines microbiological sampling. Encyclopedia of Environmental Microbiology. Wiley, New York

Murray AE, Hollibaugh JT, Orrego C (1996) Phylogenetic compositions of bacterioplankton from two California estuaries compared by denaturing gradient gel electrophoresis of 16S rDNA fragments. Appl Environ Microbiol 62(7):2676–2680

Muyzer G, Smalla K (1998) Application of denaturing gradient gel electrophoresis (DGGE) and temperature gradient gel electrophoresis (TGGE) in microbial ecology. Antonie Van Leeuwenhoek 73(1):127–141

Muyzer G, de Waal EC, Uitterlinden AG (1993) Profiling of complex microbial populations by denaturing gradient gel electrophoresis analysis of polymerase chain reaction-amplified genes coding for 16S rRNA. Appl Environ Microbiol 59(3):695–700

Nakaew N, Pathom-aree W, Lumyong S (2009a) First record of the isolation. Identification and biological activity of a new strain of spirillospora albida from thai cave Soil. Actinomycetologica 23(1):1–7

Nakaew N, Pathom-aree W, Lumyong S (2009b) Generic diversity of rare actinomycetes from thai cave soils and their possible use as new bioactive compounds. Actinomycetologica 23(2):21–26

Northup DE, Lavoie KH (2001) Geomicrobiology of caves: a review. Geomicrobiol J 18(3):199–222

Northup DE, Welbourn WC (1997) Life in the twilight zone: lava tube ecology. New Mexico Bur Mines Miner Resour Bull 156:69–82

Northup DE, Barnes SM, Yu LE et al (2003) Diverse microbial communities inhabiting ferromanganese deposits in lechuguilla and spider caves. Environ Microbiol 5:1071–1086

Northup DE, Connolly CA, Trent A et al (2004) The Nature of bacterial communities in four windows cave, El Malpais National Monument, New Mexico, USA. AMCS Bull 19:119–125

Northup DE, Melim LA, Spilde MN et al (2011) Lava cave microbial communities within mats and secondary mineral deposits: implications for life detection on other planets. Astrobiology 11(7):601–618

Onac BP, Forti P (2011) Minerogenetic mechanisms occurring in the cave environment: an overview. Int J Speleol 40(2):79–98

Pace NR (1997) A molecular view of microbial diversity and the biosphere. Science 276:734–740

Palmer AN (2007) Cave geology and speleogenesis over the past 65 years: Roles of the national speleological society in advancing the science. J Cave Karst Stud 69(1):3–12

Pankratov TA, Serkebaeva YM, Kulichevskaya IS et al (2008) Substrate-induced growth and isolation of Acidobacteria from acidic sphagnum peat. ISME J 2:551–560

Pedersen K (2000) Exploration of deep intraterrestrial microbial life: current perspectives. FEMS Microbiol Lett 185(1):9–16

Priscu JC, Adams EE, Lyons WB et al (1999) Geomicrobiology of subglacial ice above Lake Vostok, Antarctica. Science 286(5447):2141–2144

Purdy KJ, Embley TM, Takii S et al (1996) Rapid extraction of DNA and rRNA from sediments by a novel hydroxyapatite spin-column method. Appl Environ Microbiol 62:3905–3907

Rastogi G, Sani RK (2011) Molecular techniques to assess microbial community structure, function, and dynamics in the environment. In: Ahmad I et al (eds) Microbes and microbial technology; agricultural and environmental applications. Springer Science+Business Media, LLC., New York, pp 29–57, DOI 10.1007/978-1-4419-7391-5_2

Rheims H, Rainey FA, Stackebrandt E (1996) A molecular approach to search for diversity among bacteria in the environment. J Ind Microbiol 17:159–169

Robe P, Nalin R, Capellano C, Vogel T, Simonet P (2003) Extraction of DNA from soil. Euro J Soil Biol 39(4):183–190

Rohwerder T, Sand W, Lascu C (2003) Preliminary evidence for a sulphur cycle in movile cave, Romania. Acta Biotechnol 23(1):101–107

Rondon MR, Goodman RM, Handelsman J (1999) The Earth's bounty: assessing and accessing soil microbial diversity. Trends Biotechnol 17(10):403–409

Rule D, Sadoway T, Moote P et al (2011) Cures from caves: cave microbiomes and their potential for drug discovery. Oral presentation presented at the 111th American Society for microbiology general meeting, New Orleans, LA, 21–24 May 2011

Rusu A, Hillebrand A, Persoiu A (2011) Biodiversity of microorganisms in perennial ice deposits from Scarisoara ice cave (Romania). First international planetary caves workshop: implications for astrobiology, climate, detection, and exploration, Carlsbad, New Mexico, 25–28 Oct 2011. LPI contribution no. 1640, p 37

Sadoway T, Cheeptham N (2011) Susceptibility of three drug–resistant, gram–negative pathogens to antimicrobial compounds produced by cave actinomycetes. Poster presented at the 111th American Society for microbiology general meeting, New Orleans, LA, 21–24 May 2011

Sărbu SM (1991) The unusual fauna of a cave with thermomineral waters containing hydrogen sulfide from southern Dobrogea Romania. Mémoir Biospéol 17:191–196

Sărbu SM, Popa R (1992) A unique chemoautotrophically based cave ecosystem. In: Camacho AI (ed) The natural history of biospeleology, Museo Nacional de Ciencias Naturales. Consejo Superior de Investigaciones Científicas, Madrid, pp 637–666

Sărbu SM, Kane TC, Kinkle BK (1996) A chemoautotrophically based cave ecosystem. Science 272(5270):1953–1955

Schabereiter-Gurtner C, Lubitz W, Rölleke S (2003) Application of broad-range 16S rRNA PCR amplification and DGGE fingerprinting for detection of tick-infecting bacteria. J Microbiol Methods 52(2):251–260

Simon KS, Benfield EF, Macko SA (2003) Food web structure and the role of epilithic biofilms in cave streams. Ecology 84:2395–2406

Slabbinck B, De Baets B, Dawyndt P et al (2009) Towards large-scale FAME-based bacterial species identification using machine learning techniques. Syst Appl Microbiol 32(3):163–176

Smalla K (2004) Microorganisms culture-independent microbiology. In: Bull TA (ed) Microbial diversity and bioprospecting. ASM, Washington DC, pp 88–99

Snider JR, Goin C, Miller R et al (2009) Ultraviolet radiation sensibility in cave bacteria: evidence of adaptation to the subsurface? Int J Speleol 38(1):13–22

Spear JR, Barton HA, Robertson CE et al (2007) Microbial community biofabrics in a geothermal mine adit. Appl Environ Microbiol 73(19):6172–6180

Storrie-Lombardi MC, Sattler B (2009) Laser-induced fluorescence emission (L.I.F.E.): in situ nondestructive detection of microbial life in the ice covers of Antarctica Lakes. Astrobiology 9(7):659–671

Storrie-Lombardi MC, Muller JP, Fisk MR et al (2009) Laser-induced fluorescence emission (L.I.F.E.): searching for mars organics with a UV-enhanced PanCam. Astrobiology 9(7):953–964

Sugita T, Kikuchi K, Makimura K et al (2005) Trichosporon species isolated from guano samples obtained from bat-inhabited caves in Japan. Appl Environ Microbiol 71(11):7626–7629

Takada Hoshino Y, Matsumoto N (2005) Skim milk drastically improves the efficacy of DNA extraction from andisol, a vulcanic ash soil. Jpn Agri Res Quat 39:247–252

Thorseth IH, Torsvik T, Torsvik V, Daae FL, Pedersen RB, Party K-S (2001) Diversity of life in ocean floor basalt. Earth Planet Sci Lett 194:31–37

Tiedje JM, Asuming-Brempong S, Nüsslein K (1999) Opening the black box of soil microbial diversity. Applied Soil Ecology 13(2):109–122

Torsvik V, Øvreås L (2002) Microbial diversity and function in soil: from genes to ecosystems. Curr Opin Microbiol 5(3):240–245

Torsvik V, Goksøyr J, Daae FL (1990) High diversity in DNA of soil bacteria. Appl Environ Microbiol 56(3):782–787

Torsvik V, Daae FL, Sandaa RA et al (1998) Novel techniques for analysing microbial diversity in natural and perturbed environments. J Biotechnol 64(1):53–62

Vartoukian SR, Palmer RM, Wade WG (2010) Strategies for culture of 'unculturable' bacteria. FEMS Microbiol Lett 309(1):1–7

Volossiouk T, Robb EJ, Nazar RN (1995) Direct DNA extraction for PCR–mediated assays of soil organisms. Appl Environ Microbiol 61:3972–3976

Wade BD, Garcia-Pichel F (2003) Evaluation of DNA extraction methods for molecular analyses of microbial communities in Modern calcareous microbialites. Geomicrobiol J 20:549–561

Walker JJ, Pace NR (2007) Phylogenetic composition of rocky mountain endolithic microbial ecosystems. Appl Environ Microbiol 73(11):3497–3504

Walker JJ, Spear JR, Pace NR (2005) Geobiology of a microbial endolithic community in the Yellowstone geothermal environment. Nature 434:1011–1014

Ward DM, Weller R, Bateson MM (1990) 16S rRNA sequences reveal numerous uncultured microorganisms in a natural community. Nature 345(6270):63–65

Weidler GW, Dornmayr-Pfaffenhuemer M, Gerbl FW et al (2007) Communities of archaea and bacteria in a subsurface radioactive thermal spring in the Austrian Central Alps, and evidence of ammonia-oxidizing crenarchaeota. Appl Environ Microbiol 73(1):259–270

Willerslev E, Hansen AJ, Poinar HN (2004) Isolation of nucleic acids and cultures from fossil ice and permafrost. Trends Ecol Evol 19:141–147

Yakimov MM, Giuliano L, Crisafi E et al (2002) Microbial community of a saline mud volcano at *San Biagio–Belpasso, Mt. Etna* (Italy). Environ Microbiol 4:256

Yücel S, Yamaç M (2010) Selection of streptomyces isolates from Turkish karstic caves against antibiotic resistant microorganisms. Pak J Pharm Sci 23(1):1–6

Zhou J, Gu Y, Zou C et al (2007) Phylogenetic diversity of bacteria in an earth-cave in Guizhou province, southwest of China. J Microbiol 45(2):105–112

Chapter 2
Cave Biofilms and Their Potential for Novel Antibiotic Discovery

Maria de Lurdes N. Enes Dapkevicius

Biofilms in Caves

Caves often harbor extensive colorful patches that contribute to their attractiveness for cavers, such as those that can be seen in Fig. 2.1. These colorful features are natural biofilms known as bacterial mats.

Biofilms are cell populations that irreversibly adhere to solid, biotic, or abiotic surfaces and that are surrounded by an exopolysaccharide (EPS) matrix, as illustrated by Fig. 2.2. In biofilms, microbial cells live in structured, coordinated, functional communities and are able to thrive more efficiently than planktonic cells in many natural environments (O'Toole et al. 2000).

Biofilm formation starts with the attachment and immobilization of free-living (planktonic) cells, the primary colonizers, to a surface. Under adequate conditions, primary colonizers grow and multiply and cell–cell interactions are established, leading to the formation of microcolonies and the segregation of EPS. Secondary colonizers gain access to the community and the biofilm maturation process begins. Mature biofilms can shed cells to the surrounding environment, by surface erosion, sloughing off, abrasion, or predator attacks (Ghigo 2003). Fig. 2.3 summarizes the stages involved in biofilm development.

Biofilm formation is regulated by quorum-sensing mechanisms, in which bacterial cells produce, detect, and respond to small, diffusible, signal molecules (Li and Tian 2012). This cell-to-cell communication system ensures the modulation of microbial behavior in communities, altering gene expression (Davies 2006). The ability to send, receive, and process information endows unicellular organisms with the capacity to act as multicell entities, increasing their chances to survive in complex environments (Hooshangi and Bentley 2008).

M.d.L.N.E. Dapkevicius (✉)
CITA-A, Department of Agricultural Sciences, University of the Azores,
Angra do Heroísmo, Azores, Portugal
e-mail: mariaenes@uac.pt

N. Cheeptham (ed.), *Cave Microbiomes: A Novel Resource for Drug Discovery*,
SpringerBriefs in Microbiology 1, DOI 10.1007/978-1-4614-5206-5_2,
© Naowarat Cheeptham 2013

35

Fig. 2.1 Extensive bacterial mats on the walls and ceiling of Gruta da Terra Mole, a lava tube located in Terceira Island (Azores, Portugal). Photo: courtesy of Pedro Cardoso

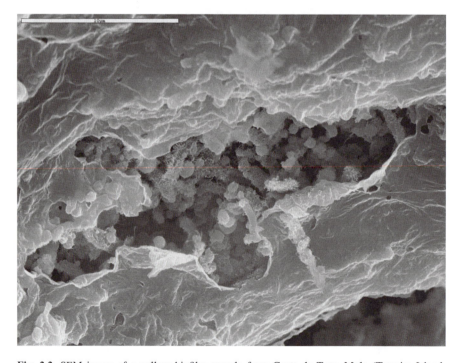

Fig. 2.2 SEM image of a yellow biofilm sample from Gruta da Terra Mole (Terceira Island, Azores, Portugal). A tear in the EPS reveals the microbial cells beneath. Courtesy of Diana Northup and Mike Spilde

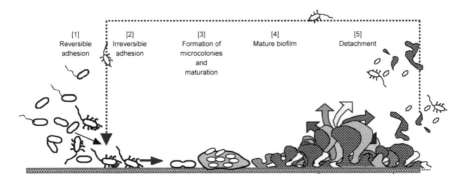

Fig. 2.3 Schematic representation of the stages in biofilm development. Reprinted from Ghigo (2003) with permission

Living in a biofilm confers adaptive advantages to the participating microorganisms. Due to the presence of the EPS, the diffusion of toxic compounds from the surroundings is restricted and nutrient adsorption is favored, leading to higher concentration of nutrients in the liquid–biofilm interface. The EPS matrix protects the cells in the biofilm from aggression by detrimental environment factors, such as UV, sudden pH variations, osmotic shock, and dehydration. The proximity among cells in biofilm communities facilitates the exchange of genetic material among them, enhancing adaptation to the environment. Close proximity in the biofilm also facilitates the development of microconsortia that allow for the utilization of renitent substrates and the colonization extreme habitats (Hall-Stoodley et al. 2004).

Extreme Environments as Sources of Useful Compounds

Extreme habitats are characterized by harsh environmental conditions and require from the organisms that dwell in them costly physiological adaptations that are absent when cells live in planktonic mode (Plath et al. 2007). The extreme environments have been defined as habitats that experience steady or fluctuating exposure to one or more environmental factors, such as salinity, osmolality, desiccation, UV radiation, barometric pressure, pH, temperature, nutrient limitation, and trophic dependence on surface environments (Boston et al. 2001; Seufferheld et al. 2008). Despite being relatively protected from factors that, in surface environments, have negative impacts on microbial life (seasonal climate variations, extreme weather, desiccation, predation by other organisms and UV radiation) (Leveillé and Datta 2010), caves are regarded as extreme habitats due to the absence of light and nutrient limitation.

Extreme environments are deemed one of the most promising sources of useful compounds. Several studies have focused on screening secondary metabolites produced by microorganisms that inhabit such environments as potential sources of useful compounds: extremozymes (Singh et al. 2011), exopolysaccharides

(Nicolaus et al. 2010), biosurfactants (Banat et al. 2010), antitumorals (Chang et al. 2011), radiation-protective drugs (Singh and Gabani 2011), antibiotics, immunosuppressants, and statins (Harvey 2000). A term has recently been coined to designate the range of biologically active, low-molecular mass compounds that are produced by bacteria, yeast, plants, and other organisms—the parvome (Davies 2009). Only a few of these molecules have already been isolated, identified, and used as therapeutic agents (Allen et al. 2010).

Antibiotics are a group of anti-infective therapeutic agents, derived from microbial sources that are used to treat bacterial infections. Antibiotics were first defined by Waksman (1945), as "chemical substances of microbial origin that possess antibiotic powers." More recent definitions do not include microbial origin, but are similar to Waksman's in that they are essentially anthropocentric, describing only a specific function (antimicrobial activity) of interest to mankind, not the chemical nature of the molecules that display it, nor their role in natural microbial populations. However, the fact that a particular metabolite behaves as an antimicrobial in the laboratory does not mean it exclusively plays the role of a "chemical weapon" in nature. This antagonistic role has been proven only in certain instances (Davies 2009). Antibiotics have a number of other effects on bacterial physiology, such as affecting the ability to swarm, the capacity to form biofilms, or inducing bacterial lysogens to enter the lytic cycle. Research carried out at the University of British Columbia, Canada, has suggested that the so-called antibiotics act as intercellular signaling agents (Davies 2009). It has been shown that all of these compounds, at very low concentrations (up to 1/100 of their inhibitory concentrations), do not exhibit antimicrobial activity, but rather display other major effects on bacterial metabolism. In the presence of such low antibiotic concentrations, the cells' global transcription patterns suffer considerable alterations. About 5–10% of cell transcripts are modulated, with approximately a half of them upregulated and the other half downregulated. The phenomenon of distinct concentration-dependent activities has been described as hormesis (Davies 2006).

From the point of view of the producing microbes, antibiotics are secondary metabolites. Like other secondary metabolites, they are not essential for vegetative growth of their producers and are synthesized from one or more primary metabolites by a wider variety of metabolic pathways than those involved in primary metabolism. Besides their antimicrobial activity, secondary metabolites may act as toxins, ionophores, bio-regulators, and signal molecules (Zengler et al. 2005).

Resistance to Antibiotics as a Driving Force for the Search for Novel Compounds

Natural product discovery has provided mankind with one of the most successful avenues for the finding of useful molecules. Life in modern human societies relies on a wide range of chemical compounds to enable economic activities and maintain adequate levels of public health protection. Natural microbiomes have provided templates for industrial enzymes as well as various types of drugs to help managing diseases with high prevalence and significance from the public health point of view, such as cancer or infectious diseases.

A 100 years ago, infections were a common cause of death and medical resources to treat them were practically inexistent. The antibiotic era in clinical medicine started more than 70 years ago with the introduction of sulfonamide. Presently, antibiotics are a highly valuable therapeutic tool, not only because mankind completely depends on them for the treatment of infectious diseases, but also because they are fundamental for the success of advanced medical procedures, such as organ transplantation, prosthesis implantations, major surgery, care for preterm babies, and cancer chemotherapy. Mass production of penicillin, which started during World War II, was a major breakthrough in the management of infectious diseases and resulted in a drastic reduction of mortality rates (Alvan et al. 2011). Because they allow for controlling a potential major cause of death among humans (infections) and have fostered considerable progress in medicine, antibiotics are regarded as one of the most successful chemotherapeutic agents. They are still extremely important in human and veterinary medicine. However, fears and concerns about their efficacy have emerged with the acknowledgment of bacterial resistance to antibiotics by the medical and scientific communities.

Antibiotic resistance was acknowledged early in the history of these chemotherapeutic agents. Fig. 2.4 shows a comparative timeline of antibiotic deployment and the appearance of resistant pathogens. Resistance to sulfonamides was reported still in the early 1940s and Fleming had already cautioned against resistance to penicillin in his Nobel Prize lecture, in 1945 (Fleming 1945). The first hospital isolate of a penicillin-resistant bacterium, a *Staphylococcus aureus*, was described in 1947, only a couple of years after the introduction in the market of this antibiotic (Ruimy et al. 2012). Presently, resistance affects all known classes of antibiotics and has attained worrisome proportions, mirrored in titles such as "Resistance to Antibiotics: Are We in the Post-Antibiotic Era?" (Alanis 2005), "Has the era of untreatable infections arrived?" (Livermore 2009), "Critical shortage of new antibiotics in development against multidrug-resistant bacteria—Time to react is now" (Moran et al. 2011), or even "Bad bugs, no drugs" (IDSA 2004). Concerned with the potential threat this situation poses in terms of public health, the World Health Organization has recently published a list of critically important antimicrobials for human medicine (AGISAR 2009).

Many of the bacterial pathogens associated with human disease have now evolved multidrug-resistant forms that are in some cases termed "superbugs," due to enhanced morbidity and mortality. Examples of these include multidrug-resistant *Mycobacterium tuberculosis, Acinetobacter baumannii, Burkholderia cepacia, Campylobacter jejuni, Citrobacter freundii, Clostridium difficile, Enterobacter* spp., *Enterococcus faecium, Enterococcus faecalis, Escherichia coli, Haemophilus influenzae, Klebsiella pneumoniae, Proteus mirabilis, Pseudomonas aeruginosa, Salmonella* spp., *Serratia* spp., *S. aureus, Staphylococcus epidermidis, Stenotrophomonas maltophilia*, and *Streptococcus pneumoniae*. Therapeutic options for these pathogens are reduced, hospitalization periods are extended, and treatments are costly. In some cases, these bacterial strains have also acquired increased virulence and higher transmissibility (Davies and Davies 2010).

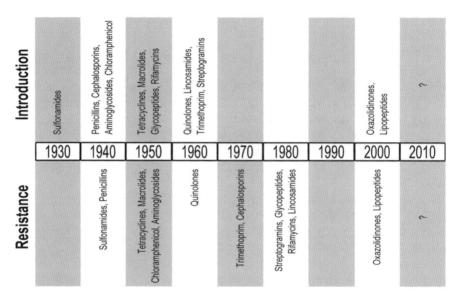

Fig. 2.4 Timeline showing the introduction of antibiotic groups and the first reports of resistance

Furthermore, novel antibiotic discovery has stalled, with no new classes of compounds added to the chemotherapeutic repertoire since the pleuromutilins, oxazolidinones, and cyclic lipopeptides. These agents, none of which is active against gram-negative pathogens, were first reported or patented in 1952, 1978, and 1987, respectively, although they were only recently (2000–2007) introduced in the market. In the golden age of antibiotic discovery, between 1940 and 1962, more than 20 new classes of antibiotics were marketed and analog development kept pace with the emergence of resistant pathogenic bacteria until one or two decades ago, but presently not enough analogues are reaching the market to keep up with the increase in resistance, particularly among gram-negative bacteria (Silver 2011). It has been estimated that, presently, at least 25,000 patients in Europe die per year because their bacterial infections are no longer treatable with the available antibiotics (ECDC/EMEA 2009). Thus, research aimed at the discovery of novel chemotherapeutic antibacterial agents is of utmost importance and urgency. Acknowledgement of this fact has led the Infectious Diseases Society of America to launch, in 2010, its 10×20 initiative, pursuing a global commitment to develop 10 new antibacterial drugs by 2020 (IDSA 2010).

Can Cave Biofilms Provide Us with Novel Antibiotics?

The first antibiotics were synthetically obtained chemicals (e.g., Salvarsan, discovered by Ehrlich in the early years of the twentieth century, and sulfonamide, developed by Domagk in the 1930s). However, the discovery of penicillin in the 1940s turned the focus of scientific activity towards natural products of microbial origin as a promising source of antibiotics. At present, the majority of successful chemical scaffolds that are still in use for this purpose are indeed dominated by microbial secondary metabolites (Wright 2011).

The dominant growth mode of bacteria is in biofilms attached to surfaces. Growth conditions in biofilms are usually heterogeneous: pH gradients may be present and physical constraints to nutrient diffusion within the biofilm may lead to differential starvation of the cells. As a result, bacterial metabolism in biofilms, including the production of secondary metabolites, is very different from what takes places in the planktonic mode (Yan et al. 2003). The production of antimicrobial molecules is one of the phenotypic characters that may be induced by the biofilm lifestyle (Rao et al. 2005). Secretion of antimicrobials aimed at inhibiting or inactivating other microorganisms is one of the main competitiveness factors within bacterial populations. It results in a toxicity gradient that precludes the growth of sensitive species in the immediate vicinity of the producer. Since the production of antimicrobial compounds can be a metabolic burden to the cell, some bacterial species evolved to start the production of these compounds only when their populations contain enough cells to accumulate an effective concentration. High enough cell densities are easily reached in biofilm communities (Moons et al. 2009). Indeed, several studies indicate that the incidence of antibiotic producing microbial isolates is higher in biofilms than in environments where the planktonic lifestyle prevails. Thus, natural biofilms have been recognized as a promising source of bacterial isolates that produce antimicrobial compounds (Mearns-Spragg et al. 1998). For instance, in marine environments, a much higher proportion of bacterial isolates that produce antimicrobials has been found in association with biofilms than in free-living bacterial populations (Burgess et al. 1999). Furthermore, Yan et al. (2003) have noted that, in certain cases, the biofilm mode of life may be indispensable for antimicrobial activity and that, when grown in laboratorial, planktonic culture isolates may stop producing antimicrobial compounds. When studying bacterial isolates from Azorean lava tubes, our work group has also observed the loss of antibacterial activity upon cultivation in liquid media (unpublished data).

Access to truly novel biodiversity has been identified as a key issue in the development of novel drugs from natural products (Zengler et al. 2005). Cave biofilms are relatively unstudied ecosystems (Boston et al. 2001) and may, thus, be promising sources for natural product research.

The physiological exploitation of the so-called "metabolically talented" microorganisms is another possible route to probe the microbial parvome for useful antibacterial chemotherapeutic agents (Zhang 2005). *Streptomyces*, some of which are able to produce a variety of secondary metabolites from different chemical families, are among the best known talented microorganisms (Knight et al. 2003).

Biodiversity in cave microbiomes is discussed in detail elsewhere in this manuscript. Although the microbial biodiversity in cave biofilms is still underexplored, several of the bacterial groups that have been described in these environments—actinobacteria, myxobacteria, pseudomonads, and *Bacillus* spp. (Menne 1999; Cañaveras et al. 2001; Northup et al. 2011)—are among the best known sources of natural products, including antibiotics.

Strategies to Increase the Chances of Success in Novel Antibiotic Discovery from Cave Microbiomes

It is generally accepted that the microbial parvome holds promise for the discovery of novel antibacterial drugs (Harvey 2000; Davies 2011; Roemer et al. 2012). However, the lack of truly novel chemicals in the antibiotic development pipeline shows that it is not an easy or a fast track to follow. Several authors have reviewed the caveats one must bear in mind when going into the path of natural product discovery (e.g., Harvey 2000; Knight et al. 2003; Zhang 2005). It is beyond the scope of the present publication to discuss them in depth. The aim of the present section is solely to raise some of the most important pitfalls that may be encountered and the developments that may help streamlining antibiotic discovery procedures and increase the possibility of a "hit."

Although the potential of well-known microbial sources of antibiotics, such as the actinomycetes, may be far from exhausted (Knight et al. 2003), the need for focusing on truly novel biodiversity in order to find compounds that will help circumvent the problem of antibacterial resistance has been emphasized (Zhang 2005). The rationale is that it will be easier to find compounds that have completely new molecular structures when searching the parvome of novel microbes. New antibacterial structures are essential, since iterations over existing ones have been exploited almost to exhaustion, and it is extremely unlikely that they will yield new, safe and effective antibiotics to fight resistant infections (Alanis 2005).

Extreme environments, biodiversity hotspots, and other underexploited habitats will most certainly provide novel biodiversity and may increase the chances of finding novel chemical structures for antibiotic development (Knight et al. 2003). Several ecosystems have been exploited in the search for novel antibacterial chemotherapeutic agents. The relatively low rate of success these efforts have met in the last few decades can be explained, in part, by the lack of systematic exploitation of the studied environments, which has led to random sampling and missed the true potential of these ecosystems (Harvey 2000).

One of the problems universally encountered when sampling for novel microbial biodiversity, especially in underexploited environments, is the low rate of microbial culturability. It has been estimated that only 0.1–1% of the microorganisms that inhabit the biosphere have been cultivated in the laboratory. Culturable microorganisms not only represent a small proportion of what can be found in nature, but possibly they are not even the most prevalent in natural populations. Increasing the proportion of culturable microorganisms is an area of research that will certainly see an increase in activity during the next years (Zhang 2005).

Cell damage by oxidative stress, formation of viable, non-culturable cells, inhibition by the high-nutrient concentrations present in current laboratory media, induction of lysogenic phages upon starvation, and lack of cell-to-cell communication under laboratory cultivation conditions have been pointed as the reasons for the low culturability rates that are presently obtained (Knight et al. 2003). Adding cell signaling molecules and using low-nutrient media have met some success in increasing the proportion of recovered microbial biodiversity (Zhang 2005). Usage of copiotrophic media and conventional culture techniques (enrichment cultures, plating out) may not only fail to provide less-culturable microorganisms with adequate conditions, but will also foster the overgrowth of fast-growing, well-known microbes (Zengler et al. 2002). Novel, high-throughput, high efficiency techniques, such as that proposed by Zengler et al. (2002), have been developed to deal with this problem. Another approach to access unculturable microorganisms' genomic information is to clone their DNA directly (Zhang 2005).

Even when performed in a systematic manner, sampling efforts will lead to an enormous amount of microbial isolates or extracts from cultures thereof. Many of these may represent re-isolation of the same microorganism/compound, making dereplication an essential part of any natural product discovery program (Zhang 2005). Dereplication will eliminate rediscoveries and, thus, contribute to more efficient, time- and cost-effective antibiotic research lines.

Isolation and characterization of bioactive compounds from natural samples is also a time- and resource-consuming task. Scaling-up the production of the candidate compound(s) in order to obtain adequate amounts for drug profiling is equally challenging. Using high-throughput methodologies that have been developed in the last decade can make this R&D step more efficient (Koehn 2008).

Summary

Microbial mats in caves are still relatively underexploited microbiomes, which may prove to contain novel biodiversity. The parvome of cave mat dwelling microorganisms could, thus, be a promising natural source of novel antimicrobial molecules that could provide truly new scaffolds for therapeutic antibiotics. Natural product discovery is not an easy endeavor, however, and well-thought, high-throughput strategies are required in order to increase the chances for success.

References

AGISAR (2009) Critically important antimicrobials for human medicine, 2nd rev report. World Health Organization, Copenhagen, Denmark

Alanis AJ (2005) Resistance to antibiotics: are we in the post-antibiotic era? Arch Med Res 36:697–705

Allen HK, Donato J, Wang HH, Cloud-Hansen KA, Davies JD, Handelsman J (2010) Call of the wild: antibiotic resistance genes in natural environments. Nat Rev Microbiol 8:251–259

Alvan G, Edlund C, Heddini A (2011) The global need for effective antibiotics—a summary of plenary presentations. Drug Resist Updates 14:70–76

Banat IM, Franzetti A, Gandolfi I, Bestetti G, Martinotti MG, Fracchia L, Smyth TJ, Marchant R (2010) Microbial biosurfactants production, applications and future potential. Appl Microbiol Biotechnol 87:427–444

Boston PJ, Spilde MN, Northup DE, Melim LA, Soroka DS, Kleina LG, Lavoie KH, Hose LD, Mallory LM, Dahm CN, Crossey LJ, Schelble RT (2001) Cave biosignature suites: microbes, minerals, and mars. Astrobiology 1:25–54

Burgess JG, Jordan EM, Bregu M, Mearns-Spragg A, Boyd KG (1999) Microbial antagonism: a neglected avenue of natural products research. J Biotechnol 70:27–32

Cañaveras JC, Sanchez-Moral S, Soler V, Saiz-Jimenez C (2001) Microorganisms and microbially induced fabrics in cave walls. Geomicrobiol J 18:223–240

Chang C-C, Chen WC, Ho T-F, Wu H-S, Wei Y-H (2011) Development of natural anti-tumor drugs by microorganisms. J Biosci Bioeng 111:501–511

Davies J (2006) Are antibiotics naturally antibiotics? J Ind Microbiol Biotechnol 33:496–499

Davies J (2009) Darwin and microbiomes. EMBO Rep 10:805

Davies J (2011) How to discover new antibiotics: harvesting the parvome. Curr Opin Chem Biol 15:5–10

Davies J, Davies D (2010) Origins and evolution of antibiotic resistance. Microbiol Mol Biol Rev 74:417–433

ECDC/EMEA (2009) The bacterial challenge: time to react. A call to narrow the gap between multidrug-resistant bacteria in the EU and the development of new antibacterial agents. Technical Report. European Centre for Disease Prevention and Control, European Medicines Agency, European Union

Fleming A (1945) Penicillin: Nobel lecture. Nobel Institute, Oslo

Ghigo JM (2003) Are there biofilm-specific physiological pathways beyond a reasonable doubt? Res Microbiol 154:1–8

Hall-Stoodley L, Costerton JW, Stoodley P (2004) Bacterial biofilms: from the natural environment to infectious diseases. Nat Rev Microbiol 2:95–108

Harvey A (2000) Strategies for discovering drugs from previously unexplored natural products. Drug Discov Today 5:294–300

Hooshangi S, Bentley WE (2008) From unicellular properties to multicellular behavior: bacteria quorum sensing circuitry and applications. Curr Opin Biotechnol 19:550–555

IDSA (2004) Bad bugs, no drugs: as antibiotic discovery stagnates … a public health crisis brews. Infectious Diseases Society of America, Alexandria, VA

IDSA (2010) The 10×20 Initiative: Pursuing a Global Commitment to Develop 10 New Antibacterial Drugs by 2020. Clin Infect Dis 50:1081–1083

Knight V, Sanglier J-J, DiTullio D, Braccili S, Bonner P, Waters J, Hughes D, Zhang L (2003) Diversifying microbial natural products for drug discovery. Appl Microbiol Biotechnol 62:446–458

Koehn FE (2008) High impact technologies for natural products screening. Prog Drug Res 65:177–210

Leveillé RJ, Datta S (2010) Lava tubes and basaltic caves as astrobiological targets on Earth and Mars: a review. Planet Space Sci 58:592–598

Li Y-H, Tian X (2012) Quorum sensing and bacterial social interactions in biofilms. Sensors 12:2519–2538

Livermore DM (2009) Has the era of untreatable infections arrived? J Antimicrob Chemother 64(S1):i29–i36

Mearns-Spragg A, Bregu M, Boyd KG, Burgess JG (1998) Cross-species induction and enhancement of antimicrobial activity produced by epibiotic bacteria from marine algae and invertebrates, after exposure to terrestrial bacteria. Lett Appl Microbiol 27:142–146

Menne B (1999) Myxobacteria in cave sediments of the French Jura Mountains. Microbiol Res 154:1–8

Moons P, Michiels CW, Aertsen A (2009) Bacterial interactions in biofilms. Crit Rev Microbiol 35:157–168

Moran LF, Aronsson B, Manz C, Gyssens IC, So AD, Monnet ID, Cars O (2011) Critical shortage of new antibiotics in development against multidrug-resistant bacteria—time to react is now. Drug Resist Updates 14:118–124

Nicolaus B, Karambourova M, Oner ET (2010) Exopolysaccharides from extremophiles: from fundamentals to biotechnology. Environ Technol 31:1145–1158

Northup DE, Melim LA, Spilde MN, Hathaway JJM, Garcia MG, Moya M, Stone FD, Boston PJ, Dapkevicius MLNE, Riquelme C (2011) Lava cave microbial communities within mats and secondary mineral deposits: implications for life detection on other planets. Astrobiology 11:601–618

O'Toole G, Kaplan HB, Kolter R (2000) Biofilm formation as microbial development. Annu Rev Microbiol 54:49–79

Plath M, Tobler M, Riesch R, García de Léon FJ, Giere O, Schlupp I (2007) Survival in an extreme habitat: the roles of behaviour and energy limitation. Naturwissenschaften 94:991–996

Rao D, Webb JS, Kjelleberg S (2005) Competitive interactions in mixed-species biofilms containing the marine bacterium *Pseudoalteromonas tunicata*. Appl Environ Microbiol 71:1729–1736

Roemer T, Davies J, Giaver G, Nislow C (2012) Bugs, drugs and chemical genomics. Nat Chem Biol 8:46–56

Ruimy R, Barbier F, Lebeaux D, Ruppé E, Andremont A (2012) Nasal carriage of methicillin-resistant coagulase-negative staphylococci: a reservoir of mecA gene for Staphylococcus aureus. In: Morand S, Beaudeau F, Carabet J (eds) New frontiers of molecular epidemiology of infectious diseases. Springer, Berlin, pp 219–238

Seufferheld M, Alvarez HM, Farias ME (2008) Role of polyphosphates in microbial adaptation to extreme environments. Appl Environ Microbiol 74:5867–5874

Silver LL (2011) Challenges of antibacterial discovery. Clin Microbiol Rev 24:71–109

Singh OV, Gabani P (2011) Extremophiles: radiation resistance microbial reserves and therapeutic implications. J Appl Microbiol 110:851–861

Singh G, Bhalla A, Ralhan PK (2011) Extremophiles and extremozymes: importance in current biotechnology. ELBA Bioflux 3:46–54

Wright GD (2011) Molecular mechanisms of antibiotic resistance. Chem Commun 47:4055–4061

Yan L, Boyd KG, Adams DR, Burgess JG (2003) Biofilm-specific cross-species induction of anti-microbial compounds in Bacilli. Appl Environ Microbiol 69:3719–3727

Zengler K, Toledo G, Rappé M, Elkins J, Mathur EJ, Short JM, Keller M (2002) Cultivating the uncultured. Proc Natl Acad Sci USA 99:15681–15686

Zengler K, Paradkar A, Keller M (2005) New methods to access microbial diversity for small molecule discovery. In: Zhang L, Demain AL (eds) Natural products. Drug discovery and therapeutic medicine. Humana, Totowa, NJ, pp 275–292

Zhang L (2005) Integrated approaches for discovering novel drugs from microbial natural products. In: Zhang L, Demain AL (eds) Natural products. Drug discovery and therapeutic medicine. Humana, Totowa, NJ, pp 33–55

Chapter 3
Cave Conservation: A Canadian Caver's Perspective

Phil Whitfield

Introduction

Apprehensive but excited, I followed my two companions through the cave entrance, a body-length vertical fissure so narrow that my chest and back barely cleared the walls as I slithered through sideways. To my relief, beyond lay a wide, high passage at right angles to the entrance fissure, at the end of which a caramel-coloured slope of crystalline rock glistened in the light of our jury-rigged headlamps. Once at the foot of the slope, I saw that a gravel-floored, irregularly ceilinged passage continued alluringly into the gloom to the left, and another much smaller but crystal-lined passage appeared to extend in the same direction from higher up the slope. In places on the slope and floor were broken fragments of soda straw-like tubes of calcite, several of which I pocketed to show my family later. My friends had told me that, because this cave had been well known for many years, the pickings for such souvenirs were slim. Apparently, they and thicker stalactites had once been everywhere on the cave roof, but the best were now either long gone or too high to reach. As we ducked, crawled, slid and climbed through the muddy, rocky and flowstone-covered passageways of the cave, my original apprehension was overwhelmed by curiosity about what might lie beyond the beam of our headlights and exploration enthusiasm took over. I was 17 years old and, like my life, the cave stretched out before me, full of unknown challenges, discoveries, fears and wonders.

P. Whitfield (✉)
British Columbia Speleological Federation, Kamloops, BC, Canada
e-mail: pwhitfield@telus.net

N. Cheeptham (ed.), *Cave Microbiomes: A Novel Resource for Drug Discovery*, SpringerBriefs in Microbiology 1, DOI 10.1007/978-1-4614-5206-5_3, © Naowarat Cheeptham 2013

A challengingly low passage, Cascade Cave, Vancouver Island, BC, Canada
(Phil Whitfield photo)[1]

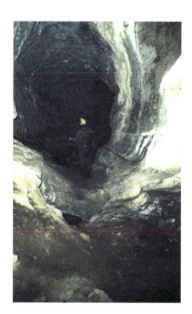

A 25 m ladder pitch in Canada's longest   Inviting "borehole" passage, Coral
cave, Castleguard Cave, Banff National   Cave, VI
Park, Alberta, Canada

[1] Unless otherwise noted, all photographs in this chapter are by the author. Vancouver Island
(henceforth abbreviated as "VI") is a large island off the coast of Canada's westernmost province,
British Columbia (henceforth abbreviated as "BC").

Six years later, I returned from postgraduate studies in England without my PhD in history, but with a Life Membership in the Cambridge University Caving Club. In the course of becoming hooked on caving, I realized how ignorant I had been on my first venture underground. The crystalline calcite soda straws I had coveted as souvenirs had been deposited, a drip at a time, at a rate probably as slow as a cubic centimetre each century. Small wonder that the rate of removal exceeded the rate of deposition!

Pristine calcite stalactites, stalagmites and "*soda straws*", Cascade Cave, VI

Breakage and mud on flowstone in Main Cave, Horne Lake Caves, VI

The mud from our clothing and boots layered the flowstone over which we climbed and crawled and had been sealed into the formations as they slowly grew, turning what were originally sparkling white crystals into the dull, caramel-coloured surfaces we had seen in every accessible part of the cave. In places where our bare hands had gripped obvious calcite handholds, the oils from our skins had interrupted the thin flow of water and blocked the deposit of crystals, leaving permanent scars. I had unwittingly become a link in what, as a cave manager, I later recognized as "the vandalism chain". The friends who took me to the original cave had themselves been taken there by their high school teacher, a member of an organized caving group, who had undoubtedly stressed the fragile and irreplaceable nature of calcite cave deposits and the importance of conserving them. However, as the third link in

Fresh mud on white calcite. Continuing crystal deposition will seal the mud into the formation

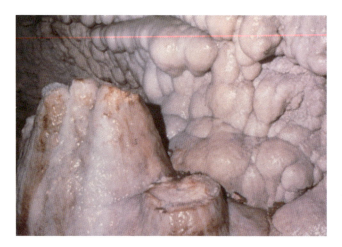

Broken and stained stalagmites, Cody Cave, BC

the chain, I missed the conservation message and thought of the cave as some kind of dark and mysterious adventure playground, probably far more dangerous to me than I could be to it. Unfortunately for cave resources, the problem of the vandalism chain remains one of the greatest challenges to cave conservation efforts.

After 3 more years and the beginning of a long career with British Columbia Parks, I became a caving club volunteer guide for members of the general public who were visiting what had been my first cave. Like me at 17, they usually arrived with no awareness of the sensitivity of the cave environment. To a very few of them, it made no difference—they were bound for adventure, not understanding. To most, however, awareness of the natural processes at work in the cave was a revelation. It was most encouraging to see small children soberly directing their parents not to touch the formations and explaining the reasons why. An educated and supportive public was obviously crucial to the success of cave resource conservation.

Beautiful crystalline cave deposits are among the most obvious of cave resources and it is usually not difficult to gain general visitor support for conserving them by not marking, breaking or walking on them. However, as I became more deeply involved with "cave management" in both my organized caving and government roles, I learned that the commonly found crystalline cave deposits were no more than an obvious tip of the iceberg in terms of sensitive cave resources.

- *Mineral resources*: Thus far, the mineral deposits found in Canadian caves are generally unexceptional by international standards, but some relatively recent discoveries elsewhere have revealed staggeringly unique crystal formations such as the massive gypsum projections of the Chandelier Ballroom in Lechuguilla Cave, New Mexico, and the enormous selenite needles of Mexico's Cueva de los Cristales at the Naica mine (refer to Lechuguilla and Naica websites for photographs).

One of Canada's most exceptional concentrations of cave mineral deposits—
helictites (eccentric stalactites) in the Chamber of Candles, Candlestick Cave, VI

Cave pearls, formed as calcite builds in layers on small pebbles constantly agitated by dripping water. A relatively rare deposit in Canadian caves, these are in Castleguard Cave

- *Paleontological and archaeological resources*: Several caves on Vancouver Island have been found to contain complete or partial skeletons of animals subsequently carbon dated to ages as great as 12,000 years. In one case, 2,600-year-old marmot bones exhibited cut markings that could only have been made by human tools (Nagorsen et al. 1996). These particular archaeological sites were

Bear skeleton, Wormhole Cave, VI. (Photo courtesy of Martin Davis)

Human skull and femur, Renaissance Cave, BC

the first of their kind on the northwest coast discovered in the mountainous subalpine region—all others had been coastal sites. Human remains found in some of these lower elevation caves have provided insights into pre- and post-contact aboriginal culture and activities.

- *Air and water*: In the course of installing gates on several high-value caves to prevent vandalism or unintentional damage, we recognized the importance of maintaining as far as possible the natural movement of air, water and animal life through the cave. Changes in air movement can affect humidity and evaporation rates, in turn changing rates of crystal deposition and the suitability of habitats for some forms of cave life. Water barriers or diversions can alter not only volumes of water entering the cave, but also the amount and type of materials that will be carried into the cave, many of which may provide food sources for cave biota. Gates blocking human access will eliminate access by any larger animals and may even prevent the passage of bats if openings are not properly spaced and aligned.

The 1971/1984 "maximum security" gate on Riverbend Cave, Horne Lake Caves Provincial Park, VI. (Photo courtesy of Rick Coles)

Inside view of the 1971 Riverbend Cave gate. Three 35 cm diameter culverts significantly reduced natural air and water movement

Riverbend Cave entrance after 2011 gate replacement. Entrance appears more natural and gate design of bars spaced 14.5 cm apart blocks unauthorized human access but allows relatively unimpeded air, water and bat movement. Photos courtesy of Rick Coles (*L*) and Lee White (*R*)

- *Sediments*: The undisturbed layers of gravel, sand, silt and mud through which we had tramped, grovelled, excavated and waded in the pursuit of virgin cave passages often told much about the surface conditions under which they had originally been carried into the caves. Large cobbles and rounded boulders suggested torrential water flows typical of glacial melting, while fine sediments tended to indicate the slowing of water movement through the cave system. In some cases, closer investigation of these sediments revealed paleontological or archaeological evidence protected from the relatively far more dynamic environment of the surface above.

Gravel seasonal streambed with transported wood fragments (Holocetus Cave, VI)

"Sandcastles" formed by dripwater in fine sand deposits (Thanksgiving Cave, VI)

- *Biota*: Some of the underground streams and pools in which we frequently had to wade, step or crawl in the course of exploration were later found to contain minute cave invertebrates, including entirely new species.[2] We may infer the presence of such obscure biota once it has been noticed, but often it may be even less obvious, and the effects of our interactions with it may be unpredictable. Our activities may inadvertently destroy a microhabitat directly by contaminating an underground pool or diverting a stream. We may find ourselves contracting histoplasmosis, a disease with symptoms not unlike tuberculosis, that is caused by inhaling the spores of the fungus *Histoplasma capsulatum*, which grows in soil and material contaminated with bird or bat droppings. Or we may be suspected of contributing to ecological disasters such as the spread of White-Nose Syndrome among bats following its 2006 outbreak in eastern North America. In that case, a fungus called *Geomyces destructans* has decimated bat populations and the possibility that cavers can unwittingly carry the spores from contaminated caves to uncontaminated ones has necessitated decontamination procedures between caving trips and triggered official closure of many caves.

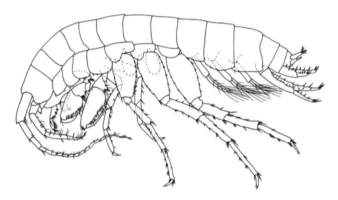

Stygobromus quatsinensis: A member of the *Crangonyctidae* family of amphipods (crustaceans with variable size and shaped limb appendages), the Quatsino cave amphipod lives exclusively in freshwater systems that flow through karst caves. There are 151 recognized *Stygobromus* species in North America. A survivor of the Pleistocene glaciation period, this species is part of a largely western group of unique freshwater invertebrates known as the "Hubbsi Group" (Drawing courtesy of D.P. Shaw, as originally submitted to the Canadian Journal of Zoology)

[2] Shaw P, Davis M, Invertebrates from Caves on Vancouver Island, Proc. Biology and Management of Species and Habitats at Risk, Kamloops, BC, 15–19 Feb 1999. http://www.env.gov.bc.ca/wld/documents/bl13shaw.pdf.

Hibernating "cave spiders"—actually not arachnids but "daddy longlegs" or harvestmen of the family *opiliones*. Often found over-wintering in Vancouver Island caves they cluster in large masses and use scent glands and "bobbing" movements to deter predators

Little brown bat with white-nose syndrome in Greeley Mine, Vermont, March 26, 2009. Credit: Marvin Moriarty/USFWS (Public domain) http://www.fws.gov/whitenosesyndrome/images/wnsGreeleyMine032609-74.jpg

WNS decontamination station at the 2009 International Congress of Speleology, Kerrville, Texas

- *Microbiota*: We understood from our basic knowledge of cave deposits or "formations" that the pasty calcite suspension known as "moonmilk" was apparently formed as microbial action somehow broke down patches of the cave walls. Such obviously fragile cave resources as moonmilk could easily be identified and treated carefully even by novice cavers. However, it took some time to realize that microorganisms, these least obvious forms of cave life, are probably present everywhere throughout the cave systems and that some of them could well be uniquely adapted to caves, or even to particular caves. We now know that some of these bacteria may have properties that would be effective in developing important new antibiotics, but it is also obvious that human intrusions might affect their existence in unforeseeable ways. Nowhere is there a better example of the dilemma facing our species than in the area of cave microbiology—we are driven to explore and discover the world around us, but we are very much in the dark about what the ultimate effects of our explorations may be, often until it is too late.

Factors such as extremely slow rates of development and structural change, absence of light, stability of temperature and relative isolation from surface systems make the low-energy environments of caves almost as distinctive and foreign as if they occurred on another planet. Indeed, it is no coincidence that NASA studies various aspects of caves to gain insights about conditions that could support extraterrestrial life. Given the unique scientific values of cave resources, cavers, consciously or not, are acting out the Jekyll and Hyde role that our species has played

"Moonmilk" deposits in Whitewall Cave, VI

A scanning electron micrograph demonstrating unidentified fossilized filamentous and rod-shaped bacteria in a volcanic cave, Wells Gray Park, BC (Photo courtesy of Dr. Naowarat Cheeptham)

on the planet from the beginning. Basic human curiosity along with the urge to explore and to seek adventure, fame or wealth led Europeans across the Atlantic and Pacific Oceans in past centuries. American and Asian cultures were discovered that proved to be more advanced in some ways than those of the explorers. Constructively, geographers like Ferdinand Magellan and David Thompson began to record and understand for the first time the relationships and layout of the continents. Naturalists like Charles Darwin were able to catalogue the global distribution of obvious plant and animal species. Less constructively, both unintentionally and intentionally, European diseases were transmitted to populations having no resistance to them, introduced species wiped out native flora and fauna and the records of unique cultures and entire civilizations were destroyed. In a similar vein, most cavers outside the small organized caving community are motivated by little more than the adventure of caving, with its physical and psychological challenges, its comradeship and the satisfaction and bragging rights associated with surviving dangers and "conquering a last frontier". Just as Cortez and Pizarro were drawn on by gold, the term "scooping booty" is applied in the caving world to those who rush to explore new passages rather than taking the time conscientiously to survey them.

The underlying drive for adventure, challenge and discovery remains an aspect of human nature, and human nature is notoriously difficult to change. Awareness of the variety, complexity and sensitivity of cave resources is a revelation for most people, but it also poses a challenge for the organized caving community. The knowledge that mere human presence underground will alter the existing balance of the cave environment argues for a precautionary and methodical approach to exploration. However, when most of those who cave actively are primarily drawn to the pastime out of a sense of adventure and physical challenge, organized caving risks alienating potential recruits and isolating itself from "recreational cavers" if conservation messages are delivered too heavy-handedly. The degree to which individuals impact cave resources ultimately hinges on their personal awareness, patience and respect for the cave environment, and these factors can be influenced by informed peer pressure, group culture and societal standards. Within the organized caving community, value is placed on cave mapping, resource inventory and specialized studies as elements of exploration—"recreation in the service of science". The legitimacy of recreational caving is recognized and supported, but recruits who initially may be interested only in the recreational aspects of caving are encouraged and mentored to achieve a better understanding of cave resources and they are urged to assume a personal stewardship role as part of their caving activities. North American cavers are guided by some version of the popular mantra: "Take nothing but photographs. Leave nothing but footprints. Kill nothing but time". Some caving organizations go farther in stating their expectations. For example, all members of the BC Speleological Federation (BCSF) are asked to sign off on a "Caver's Code of Conduct" that attempts to provide general guidelines for both cave conservation and caver safety. To foster a "corporate culture" of collective and personal responsibility, the leadership of the Federation plays an active role in public education, cave data collection, cave management planning and the direct protection of vulnerable caves with significant resource values.

Volunteers removing spray paint graffiti, Cody Caves, BC

Caving club members conducting photographic monitoring in Candlestick Cave, VI

A speleologist taking notes during paleoclimatic studies involving the carbon-dating of speleothems in a Vancouver Island cave

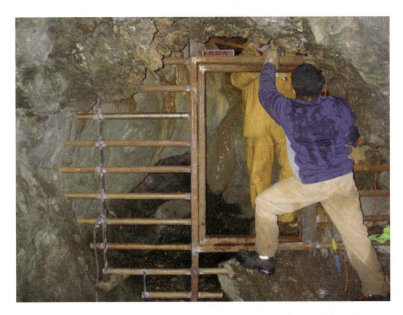

Canadian Cave Conservancy workers installing a gate on Lower Cave, Horne Lake Caves Park, VI, a first step in a long process of restoration after decades of damage

It is difficult to measure the effectiveness of this approach in terms of cave conservation. Within the organized caving community, no one can claim to be unaware of the importance of conservation, but in North America few caving organizations provide their members with structured training in cave conservation practices. Despite any "culture of conservation" and mentoring of new members, detailed knowledge of what to look for and how to minimize impacts may take some time to assimilate. In my personal experience, some caves where access has been limited to organized caving groups have escaped deliberate damage but have still experienced much avoidable wear and tear. In other cases where a single individual has been designated as a "cave custodian" who must accompany all parties visiting a controlled-access cave, impacts have been minimal. The lesson learned seems to be that trying to hold everyone collectively responsible tends to leave no one feeling responsible, whereas focusing responsibility and some authority on a single individual ensures both accountability and consistency. Accordingly, the policy of the BC Speleological Federation has been to apply or advocate focused custodianship arrangements for any high-value, controlled-access caves. Following the practice of Alaska cavers, British Columbia cavers who discover, explore and map caves are also encouraged to include simple management direction statements on their cave surveys. This approach obviously cannot ensure that anyone subsequently visiting the cave will follow the proposed directions, but it at least communicates the conservation and safety concerns initially noted. Such flagging of concerns tends to be respected by other members of the caving community, who may then build on this basic assessment of the cave's resources and perhaps add to it with their own observations and advice.

Clearly, effective communication is key to identifying and conserving sensitive cave resources. Those who join caving organizations are immediately plugged into newsletters, forums and other media that educate them about cave resource values and conservation issues. However, in North America and worldwide, the organized caving community probably includes fewer than 5–10% of the number of people actually caving. The quandary for organized caving is how to communicate conservation and safety messages to these unknown individuals without seeming actively to promote caving and thereby increase user pressures on the caves. Generally, the approach has been to ensure that any material supplied to the public media contains conservation and safety messages and channels interested recipients toward caving clubs. In addition, some caving organizations have posted contact information at commercial caves and at cave entrances that are known to be visited by casual cavers. Practically speaking, these measures tend to recruit only a relative few who are seriously interested in caves and caving, which at least ensures the survival and renewal of "organized" caving. Unfortunately, there remains a large number of freelance cavers, most of whom have little insight into the values of cave resources and a few of whom—the speleothem collectors, graffiti artists, bat bashers, etc.—may be bent on their deliberate destruction.

In the past, organized caving has been somewhat successful in minimizing the impact of general recreational caving by directing it toward "durable" caves with few known vulnerable resources or to controlled-access caves where activities are supervised.

Potentially sensitive caves have been kept under wraps. This information management approach was fairly effective in the days when someone might find out about a cave only through a personal visit, having first won the trust of a caving club after a period of face-to-face contact. Even if this initiated a vandalism chain in rare cases, the chain depended on information transfer from one individual to a relatively few other individuals and over a period of time. Location records on paper could only be disseminated by photocopying, making it easier to control who would obtain copies and how widely copies might possibly be proliferated. However, managing use by managing cave location information is now jeopardized by the advent of GPS and digital records. Anyone visiting a cave can now plot its precise location by GPS and post that information to the world via Google Earth or some similar application. Digital caving community inventory and location data on all known caves can be broadcast anywhere with a single keystroke and once the genie is out of the bottle all direct control is lost. The default position is reliance on a higher level of social awareness and societal ethics around cave conservation, which in turn requires support in law.

Nor are those who actually go underground the only threat to cave resources. As I became more deeply involved in Vancouver Island caving in the 1970s, our primary means of locating caves was to receive tips from forestry workers or to drive around recently clear-cut areas where limestone outcrops, sinkholes and cave entrances stood out obviously, no longer concealed by the dense temperate rainforest. All too often, logs, stumps and other debris had been dumped into sinkholes and small streams that flowed into cave entrances were choked with silt, branches and

Clearcutting and broadcast burning on karst causes rapid run-offs and erosion of soils into caves below (Photo courtesy of Ken Sinkiewicz)

Logging debris nearly choking a cave entrance in a sinkhole, Vancouver Island (Photo courtesy of Ken Sinkiewicz)

tree bark. In some cases, road cuts or rock quarries had opened up caves previously sealed from the surface, and in others, rock and gravel had filled in holes of any size that interfered with roadbuilding. Over time, it became clear that the characteristically heavy rainfalls on the Island were eroding the surface soils of the clear-cuts in karst areas, while the increased run-offs were significantly changing the hydrology and the biology of the caves below. Managing human activity underground was obviously only half the battle; surface activities also required regulation if cave and karst resources were to be conserved. Here, the caving community could only hope to prevail by educating and winning the support of governments and sympathetic lobbyists in the broader, non-caving society. In this, despite its small size and low (even subterranean) profile, organized caving has been remarkably successful.

Fortunately, broad societal awareness of environmental issues in the Western world has increased dramatically over the last half century, creating a more positive climate for cave conservation than existed in the days of commercial guano and speleothem mining, when sewage was diverted into caves and sinkholes were used as convenient garbage dumps.

By the 1980s, responding to the obviously legitimate concerns of the small, organized caving community, a few sympathetic staff members within the British Columbia Forest Service began to develop guidelines for timber harvesting in karst areas. Still, it was another 20 years before legislation and regulation provided any legal compulsion in support of the evolving "best management practices". Meanwhile, in the USA and other countries, cavers and environmentalists had succeeded in lobbying for legislation that would protect cave and karst resources both from damaging activities underground and from surface activities. Possibly because

Canada lacks America's innumerable private commercial caves and high-profile tourist caves such as Carlsbad, Mammoth, Wind and Jewel, Canadian caves remain well below the radar in terms of public understanding and governments have been slow to develop legislation that publicly affirms cave resource values and effectively protects them. Even if no caver ever set foot in a cave, conservation of cave resources would be far from assured.

In British Columbia, for example, the current legal framework has given cave/karst conservation issues greater profile with the forest industry. However, it applies only to forestry-related industrial activities and has no application to mining, quarrying, hydropower generation, highway construction, real estate development or any of the many other surface activities that potentially can impact cave and karst resources. Additionally, the current "results based" regulations may well be virtually impossible to enforce, as they rely on evidence of damage to cave systems that can be demonstrated conclusively only if detailed baseline conditions had been recorded before the industrial disturbance took place. While in the USA, 26 internal jurisdictions and the federal government have specific cave conservation legislation in place, Canada has none. In British Columbia, cave resources are protected by law only if they lie within Provincial Parks or formally designated Provincial Recreation Sites, are subject to damage by industrial forestry activities, or are explicitly protected under Heritage Conservation legislation, blanket provisions of which cover only archaeological resources. If legislation is seen as a means of "setting the bar" for particular societal values, British Columbia clearly has yet to recognize adequately the value of its cave resources. Protection of cave/karst resources from both industrial scale impacts and deliberate damage by underground visitors remains patchy and inadequate.

As noted previously, communication and education are key to promoting cave conservation, both inside the organized caving community and across the broader society. As cavers, we realize that, whatever we may have achieved in terms of cave resource protection, we still have much to learn and to do. This sense of responsibility can sometimes detract from the enjoyment of simply "going caving", but we understand that this enjoyment will not be sustainable if we cannot somehow preserve the resources that make caves so fascinating.

Reference

Nagorsen DW, Keddie G, Luszcz T (1996) Vancouver Island marmot bones in subalpine caves: archaeological and biological implications (Occasional Paper No. 4). Ministry of Environment, Lands and Parks, BC Parks, Victoria

Chapter 4
Cave Conservation: A Microbiologist's Perspective

Cesareo Saiz-Jimenez

Alice opened the door and found that it led into a small passage, not much larger than a rat-hole: she knelt down and looked along the passage into the loveliest garden you ever saw.

Alice's Adventures in Wonderland, Lewis Carroll

Issues on Cave Conservation

Rocks and landforms are the Earth's memory. Included among landforms are caves, formed by erosion and the weathering of rocks. The most common types of caves are found in limestone and other calcareous rocks, and in lava tubes on basaltic rock. The remaining types, including those formed in gypsum, quartzite, sandstone, and granite, are usually limited in extent and not always found and investigated.

The United Nations Educational, Scientific and Cultural Organization (UNESCO) encouraged the identification, protection, and preservation of cultural and natural heritage of outstanding value to humanity. For UNESCO, natural heritage refers, among other things, to outstanding geological formations with scientific, conservation, or aesthetic value (e.g. caves), and cultural heritage refers to sites with historical, aesthetic, archaeological, scientific, ethnological, or anthropological value (e.g. caves with rock art).

Some examples of World Heritage Sites of Geological Interest are Carlsbad Caverns National Park and Mammoth Cave National Park in USA and Aggtelek Karst and Slovak Karst in Hungary/Slovakia. World Heritage Sites of Cultural Interest are Cave of Altamira and Palaeolithic Cave Art of Northern Spain (Fig. 4.1), Prehistoric Sites and Decorated Caves of the Vézère Valley, France, Mogao Caves in China, etc. Other less important caves benefit from some national or regional protection status.

C. Saiz-Jimenez (✉)
Instituto de Recursos Naturales y Agrobiologia, Consejo Superior de Investigaciones
Cientificas (IRNAS-CSIC), Avda. Reina Mercedes, 10, 41012 Sevilla, Spain
e-mail: saiz@irnase.csic.es

N. Cheeptham (ed.), *Cave Microbiomes: A Novel Resource for Drug Discovery*,
SpringerBriefs in Microbiology 1, DOI 10.1007/978-1-4614-5206-5_4,
© Naowarat Cheeptham 2013

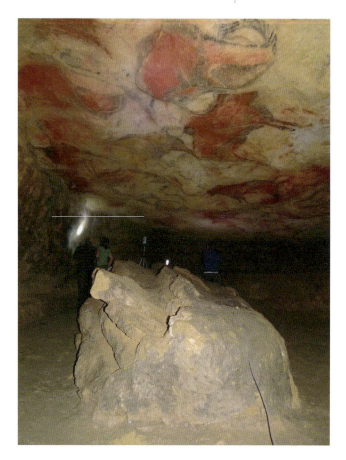

Fig. 4.1 Ceiling of the Polychrome Hall. Altamira Cave, Spain

Humans are the main threat to the conservation of a cave and in this group we must distinguish several categories, each one corresponding to different levels of responsibility: cave managers and staff, visitors, and scientists. Cave managers and staff are the people ultimately responsible for the conservation of caves, but political pressures or scientific ignorance can lead to adopting poor decisions that can mark the fate of a cave and the associated damage. Thus, an example of poor management is the construction of replicas near the original cave, which include parking lots, restaurants, etc. This contributes to air and ground pollution and attracts rodents, which find an easy nutrient source in the garbage and then seek protection in the cave. Rodent droppings are excellent nutrients for microbial growth and insects. Many times these problems can be solved by requesting advice from specialists in the conservation of caves, prior to designing a project or making a poor decision.

Tourism is one of the world's largest industries, but spawns well-known problems, because tourism activities lead to important environmental impacts caused by the adaptation of caves to visits and by the visitors themselves. Some examples of

Fig. 4.2 Alain Roussot and André Glory at Lascaux Cave, France, in 1953. They are applying a translucent medium directly on the rock surface to trace the image. Photo from http://donsmaps.com/lascaux.html

severe impacts were the construction of concrete walls, access galleries, and artificial halls in Lascaux and Altamira caves at the time of discovery or later (Geneste 2011; Lasheras et al. 2011). Tourism pressure often fostered by the political authorities, as an undue form of development of a region, results in actions that are counterproductive to the conservation of caves (Saiz-Jimenez et al. 2011). Unfortunately economic interests often prevail over the protection of the natural site and the rock art.

For example, human visitation can introduce new organic matter and exotic microorganisms into caves, which may harm native microbial populations (Northup 2009) in addition to the direct damage inflicted on the cave. The visits leave behind skin cells, bacteria and fungi from hair and skin, hair, and occasionally vomit, faeces, urine, or mud and dirt from other caves. One of the major impacts on oligotrophic (low-nutrient) caves is the enrichment of new organic carbon from outside. In some cases, caves had to be closed due to the remarkable deterioration that occurred after a large number of visits, and tourists had to be re-routed to a replica cave (Lasheras et al. 2011).

Scientists are also a threat for caves. They enter the cave with sound intentions, but, for the sake of a study or publication, natural and cultural heritage can be destroyed. Several examples are the drilling of rocks and speleothems (Spötl and Mattey 2012), the recording of paintings, or the sampling of pigments for chemical analysis or dating. In the past, paintings were recorded by archaeologists with inadequate methods such as placing transparent sheets over the painting (Bednarik 2007). The consequence of this was the removal of pigments that attached to the sheet during the recording process (Fig. 4.2). Another example is the sampling of considerable amounts of pigments for analysis. From the Upper Palaeolithic record, red ochre (hematite) is indeed well known for its use in cave paintings (Chalmin et al. 2003; Roebroeks et al. 2012), but archaeologists insist on taking samples from different rock art with red pigments, perhaps expecting they will find a different mineral. Today, with the development of portable instruments for non-destructive and contactless analysis (XRD/XRF, Raman, etc.), there is no need to collect pigment samples (Pifferi et al. 2009; Jehlicka et al. 2009).

Of special importance is the loss of pigments that many French and Spanish caves suffered for radiocarbon dating purposes. Before the 1990s huge amounts of pigment were needed for dating. Radiocarbon techniques for measuring cave art pigments have been available since the 1990s with the use of an experimental Accelerator Mass Spectrometer (AMS) in France. This technique reduced the sample size from 10 to 50 mg (Clottes et al. 1992; Valladas et al. 1992); however, this amount was enough to produce a visible damage in the paintings. Valladas (2003) states that to protect the visual integrity of the drawings, pigment is scraped from rock cracks or from the thickest layers, but this is not always true as we observed in a Spanish cave. Today, with the new instrumentation available the amount of sample needed for a radiocarbon dating has been reduced to one-tenth of the amounts commonly used in the 1990s. Currently, with the rapid development of more accurate and sophisticated instruments, the damages inflicted on the paintings in the past for the sake of having a date were unjustifiable, and the data obtained with experimental devices produced serious discrepancies, which were strongly debated by archaeologists (Züchner 2003; Pettit and Pike 2007; Pettit 2008).

At some distance from humans are the animals that inhabit the cave. In many cases they represent a clear threat to the conservation of the paintings. Rodents enrich the caves with excrement, which is a good nutrient source for bacteria, fungi, collembolans, and other small insects. Large amounts of fungi have been observed as a consequence of rodent activity in some Spanish caves (Hermosín et al. 2010; Saiz-Jimenez et al. 2011). In addition, bats contribute largely to cave pollution due to the excretion of important amounts of guano. In these cases one is facing the dilemma: protect the rock art or protect the bat colony?

A Cave Microbiologist's Perspective

A cave is a fascinating world for a microbiologist. There, one can find from the simplest to the most complex microbial ecosystems and from chemolithotrophic (e.g. methanogens) to heterotrophic microorganisms degrading the most recalcitrant macromolecules. Caves, for a microbial ecologist, are Alice's *loveliest garden*, where on a journey one can recall lessons learnt in a "*book without pictures*". These *loveliest gardens* have plenty of colourful minerals (Figs. 4.3 and 4.4) and microbial colonisations (Fig. 4.5), fragile just as cave's own ecosystem is.

Fig. 4.3 General view. Pozo Alfredo mine, Riotinto, Spain. Photo ©Manuel Aragón

Fig. 4.4 Detail of the ground. Pozo Alfredo mine, Riotinto, Spain. Photo ©Manuel Aragón

Caves are populated by microorganisms and animals (Lee et al. 2012). Because caves are extreme environments subjected to stringent conditions, cave microorganisms represent an invaluable genetic pool of distinctive species-specific characteristics. As far as *Archaea* and *Bacteria* are concerned, most of the 16S rRNA gene sequences retrieved from caves represented novel and thus far uncultivable species with unknown functions (Schabereiter-Gurtner et al. 2002a, b, 2004). For this reason,

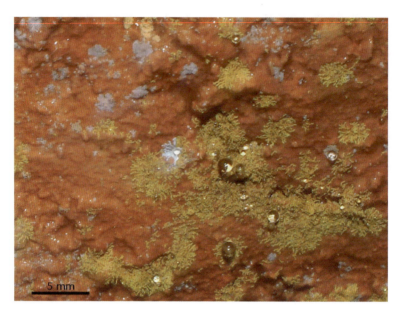

Fig. 4.5 Bacterial colonisation of a wall. Altamira Cave, Spain

caves are unexplored terrestrial ecosystems to search for novel genes and bioactive substance-producing microorganisms (Groth and Saiz-Jimenez 1999), similarly to how scientists regarded the marine environment (Debbab et al. 2010). An example is *Streptomyces tendae*, which was isolated from Grotta dei Cervi in Italy (Groth et al. 2001), and was found to produce cervimycin, a new glycoside antibiotic that is active against methicillin-resistant *Staphylococcus aureus* (Bretschneider et al. 2012).

In the past, microbiologists needed a lot of sample to isolate microorganisms. It was believed that isolated microorganisms corresponded to those growing or biodeteriorating the cultural heritage. This is not always true and it was stated that the microbial spectrum obtained using cultivation methods was completely different to that obtained using molecular tools (Laiz et al. 2003).

Experienced cave microbiologists are aware that a sound knowledge of the habitat is fundamental to understanding and interpreting the phenomena occurring in a subterranean ecosystem in a proper way. This is not always the situation. Many "laboratory-based" and molecular microbiologists, not familiar with the complex cave microbial ecology, tend to provide long lists of bacterial clones. These lists lack an overall ecological concept. A short or single visit for collecting a few samples often results in obtaining poor data and in confusing interpretations of the ecosystem. The situation is particularly worrying when erroneous interpretations appear in technical reports or papers that can cause a lot of confusion or report on inaccurate data. Taking into account that cave research and the information obtained is the key for correct cave management and conservation, the matter should be tackled

Fig. 4.6 Fungal growth on a wall. Ardales Cave, Spain (April 1995)

with extreme caution. Unfortunately, in the literature there are a few examples of erroneous approaches and inferences in cave microbiology, mainly derived from a lack of data and its incorrect interpretation, as will be discussed later. Reviewers and editors of journals are also accomplices to big mistakes by not adequately carrying out their jobs and not detecting such errors.

One of the most critical aspects in the study of cave microorganisms is the need to know this particular ecosystem and to confront the microbiological data with those of geology, mineralogy, geochemistry, hydrology, etc.; otherwise many of the characteristic features of the ecosystem may go unnoticed. A lack of information on fungal taxonomy and ecology will not recognise that many of the fungi identified in subterranean environments can be catalogued as entomogenic or entomopathogenic and a direct relationship with arthropoda may be traced (Jurado et al. 2008; Bastian et al. 2009c, 2010).

Another issue is that some microbiological studies are carried out without minimal information on the environment and their nutrient sources, thus providing erroneous data or assuming facts which have not been tested experimentally. For instance, Ardales Cave, also known as Doña Trinidad Cave, Spain, shows a marked fungal colonisation as a consequence of eutrophisation due to the anthropogenic influence on the cave during 190 years. In fact, this cave was even densely populated by the residents of Ardales village during the Spanish Civil War in 1936–1939. Frequent visits and studies in this cave in the last two decades (e.g. Saiz-Jimenez 2001; Hermosín et al. 2010) revealed the presence of fungi (Fig. 4.6), bats, and rodents, and droppings and excrement could be found everywhere including along the visitor's trail (Fig. 4.7). This is also reflected in the relatively important levels of ammonium and phosphorus determined in the chemical analyses of ground sediments.

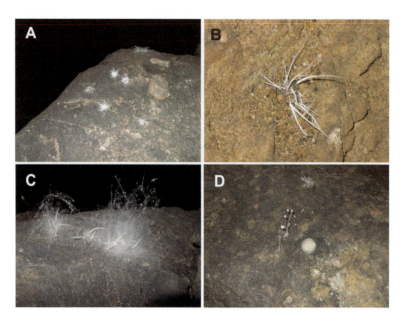

Fig. 4.7 Several types of fungi in the ground of Ardales Cave, Spain. (**a**) Abundant rodent excrements colonised by *Beauveria felina*. (**b**) Detail of *Beauveria felina* synnemata. (**c**) *Beauveria felina* and *Phycomyces nitens* on a rodent excrement. (**d**) Agaricaceous fungi on rodent excrement (March/April 2010)

These data are of importance to avoid misunderstandings concerning this cave and its nutrient limitations. In fact, Stomeo et al. (2009) inaccurately reported that in Ardales Cave "*massive fungal growth was not visible but Fusarium spp. was the only fungus that could be both isolated and detected through molecular methods*". However, Hermosín et al. (2010) proved that fungal growth in this cave has always been evident on the different visits carried out in the last few years, either on the sediments and animal excrements or by dissemination of fungal spores in the air (Fernandez-Cortes et al. 2011). The isolation of only one fungal strain of *Fusarium* by Stomeo et al. (2009) poses reasonable questions about the protocols for culturing fungi used by the authors, their samplings, and field surveys, if any. Hermosín et al. (2010) reported that 40 different taxa were isolated from this cave from different niches (dripping waters, sediments, rodent excrements, animal hairs and bones, etc.), which include among others *Fusarium solani* and *Fusarium* sp. Furthermore, a luxurious growth of many other fungi in the cave was observed (Fig. 4.7).

For Stomeo et al. (2009) "*the relatively low temperature and limitation in ammonium might be restricting the development of Fusarium in Doña Trinidad Cave*". This assertion is highly questionable. The temperature in Ardales Cave (16.3–17.6°C) is not a handicap for fungal growth as demonstrated in Figs. 4.6 and 4.7. Massive outbreaks of *F. solani* have been reported in Lascaux Cave (12.4–12.7°C) (Dupont et al. 2007; Malaurent et al. 2011) or Castañar de Ibor Cave (16.9–17.0°C) (Jurado et al. 2010).

In addition, it was observed that *F. solani* and other *Fusarium* spp. were growing over wooden materials in Ardales Cave. Therefore neither temperature nor ammonium (derived from guano and rodent excrements) is a limiting factor for the growth of fungi in this cave. Thus, the mistakes of some microbiologists can lead to the application of negative measures in the management of a cave.

Conservation Versus Restoration

Recently, it was stated that "with today's leisure tourism, the frequency of visits to many caves and other subterranean sites should be looked upon as a potential risk for the conservation of cultural heritage. Archaeologists, environmentalists and microbiologists agree on the beneficial effect of closing subterranean sites for their conservation" (Saiz-Jimenez et al. 2011). These authors reported that in Altamira Cave the 2002 closing represented a clear benefit for the conservation of the paintings; the green phototrophic colonisations did not continue to progress, and the corrosion rate of the paintings' host-rock decreased. A progressive decrease was also noted in the content of organic matter and nitrogen compounds in the infiltration waters after the correction of the human activities in the top soil affecting the cave and a control of the vegetation in the outer zone. The installation of a new access door in 2007 equipped with a thermal insulation system and the placement of a second door at a distance of 20 m from the main entrance reduced the entry of airborne particles, the condensation rate in the entrance area, and the metabolic activity of the main visible microbial colonies.

When visits cannot be prevented, adequate cave management, including a sound scientific management of the microclimate, environment, and visits, is the best way to help preservation. In some caves with pronounced disturbances, damage restoration has been undertaken. In cultural heritage, restoration refers to those specialised treatments of the materials in order to prolong the life and the information that the object carries through time, without changing its appearance. A successful rock art cave restoration and preservation is dependent on knowledge which is rarely found in the cultural heritage management authorities. Usually these authorities and cave managers have academic formation in the fields of museology, archaeology, history, or architecture. Disciplines necessary for understanding the complex functioning of caves rely on geology, hydrology, climatology, physics, chemistry, microbiology, zoology, etc. not alone or independently from one another, but integrated in a global concept. Therefore, errors are common due to the lack of interaction between different professionals and disciplines, or even when this interaction occurs, to the mistakes originated by poor advising on the part of non-specialised professionals.

In a natural state, caves are oligotrophic (nutrient poor) environments with very little connection to the outside atmosphere. Once discovered, the numerous conditioning projects for facilitating cave visits, the impact of anthropogenic activities in the top soil (agriculture, livestock), urbanisation (construction of housing, shops, restaurants, parking areas, etc.), and the massive visits transformed the pristine ecosystem into one with an abundance of available nutrients. Lavoie and Northup (2006) suggested that selected bacteria can be proposed as indicators of human activities in US caves. They include *Escherichia coli*, *S. aureus*, and *Bacillus* spp. The results showed increased levels of *S. aureus*, *E. coli*, and high-temperature *Bacillus* in areas with the greatest visitation levels in both wild caves and commercial caves.

A number of European show caves, with many years of accumulated changes, have little hope of complete restoration. Boston et al. (2006) stated that "restoration is no longer restoration when it degenerates merely to cave housecleaning. The object of restoration is to maintain the cave's natural state or to return the cave to a former natural state that existed before impact by human activities". However, for Elliott (2006) no cave is ever completely restored to its former aesthetic or ecological state once it has been ecologically disturbed. An excellent example of an ecologically disturbed cave is Lascaux in France (Bastian et al. 2010; Martin-Sanchez et al. 2012a, b).

In an excellent review Boston et al. (2006) discussed the effect of different cleaning chemicals on microbial communities and mineral formations. Interestingly, no biocides were mentioned in the review, perhaps because an experienced caver will never use such kind of products. Unfortunately, some cave restoration efforts in the last decade applied conventional biocides that are likely not suitable for the unknown and complex microbial communities growing on and beneath the rock surfaces. Restoration efforts can also have an opposite effect, particularly if the restorers use substrates that support the growth of microorganisms and, consequently, accelerate the deterioration process.

Several bioremediation approaches involving biocides have been used to prevent biodeterioration and microbial outbreaks. In the last decade and since the massive colonisation of Lascaux Cave by *F. solani* in the year 2001, a controversy has emerged with the media, associations for the protection of cultural heritage, and the French Ministry of Culture arguing over the causes of the fungal colonisation and the ways in which to combat it. Only very recently has it been possible to define the microbiology of this cave, as summarised by Bastian et al. (2010) and Martin-Sanchez et al. (2012b).

The biocide benzalkonium chloride was used for only a few years in Lascaux Cave to control the outbreak of *F. solani* in the year 2001, which was characterised by the formation of masses of white mycelia on the ground sediments and walls. However, the application of this biocide selected for bacteria (Bastian et al. 2009a, b) and fungi (Martin-Sanchez et al. 2012b) that were resistant to treatments; subsequently, black stains produced by the growth of dematiaceous fungi (whose walls contain melanin) appeared. It is well known in industrial microbiology that a rotation of biocides is required to prevent microbial resistance (Langsrud et al. 2003). Lascaux is an example of how an intensive treatment with a single biocidal product is not adequate to combat fungal outbreaks (Bastian et al. 2009a, b). However, in 2008 a biocide treatment was repeated on the black stains with a commercial product which contained a mixture of benzalkonium chloride, miristalkonium chloride, and the fungicide 2-octyl-2H-isothiazol-3-one, and additional applications of solution of isothiazoline derivatives. Martin-Sanchez et al. (2012b) have the opportunity to follow this treatment and reported that the black stains were mainly formed by the fungus *Ochroconis lascauxensis* (97% of clones). Six months after cleaning and treatment of black stains with the biocides resulted in some decrease of the main fungus involved in stain formation (from 97 to 52.3% of clones), but on the other hand, the fungal community was more diverse, and corresponded to the frequent airborne genera *Aspergillus* (13.6%), *Cladosporium* (11.4%), *Trichoderma* (9.1), and *Alternaria* (4.5%). In areas treated with biocides, when *O. lascauxensis* was partially removed and other bacteria and fungi killed, there was a rapid succession in the ecological niche, which was occupied by airborne fungal spores. They quickly colonised the treated areas as saprophytic fungi specialised in the decomposition of organic matter. Dead microbial biomass and degradation products from biocides likely trigged this secondary colonisation and conducted to an increase in the microbial diversity.

A similar fungal outbreak occurred on August 24, 2008, in Castañar de Ibor Cave, with excessive growth of *F. solani* and *Mucor circinelloides*, which was caused by an accidental discharge of organic waste (vomit). After 40 h, the massive colonisation of the floors and walls was documented and treated effectively (Jurado et al. 2010). The use of benzalkonium chloride, as was performed in Lascaux Cave, was not considered because it leads to serious environmental problems and subsequent colonisation by other fungi. A mechanical cleaning of the surface of contaminated sediments was employed. In addition, hydrogen peroxide was used to oxidise residual organic matter and eliminate inaccessible fungal structures. This method succeeded.

The hydrogen peroxide method (Faimon et al. 2003) has advantages because when it comes into contact with organic matter, it is decomposed into products that are harmless to the cave environment (water, oxygen, and CO_2). In addition, it can be easily applied by the cave personnel without a risk to their health or the need for special precautions.

Mulec and Kosi (2009) reviewed the pros and cons of physical, chemical, and biological methods to control phototrophic growth in caves. For these authors, currently there is no ideal solution. Akatova et al. (2009) reported several bioremediation techniques in an intervention proposed to clean affected surfaces in the Salpetre Cave of Collbató, Barcelona, Spain, a cave without rock art, and to prevent the growth of subsequent phototrophic biofilms. To this end, two equivalent testing areas with a relatively flat topography were selected inside the cave. The substratum was a stalactite, which was divided into four vertically arranged quadrants (Fig. 4.8).

For each area to be tested, one quadrant (A) was maintained as a control, as a reference for the possible effects of environmental change on local biofilm growth throughout the year. The three remaining quadrants (B, C, and D) were cleaned using a synthetic brush embedded in 70% alcohol/distilled water to remove photosynthetic organisms and to allow better penetration of biocidal products inside the substrata. Quadrants C and D were then treated with benzalkonium chloride, using both the commercially available active ingredient (1 ‰) and the complex commercial organic compound (Preventol R50, 2%). In addition, the effectiveness of the treatments was evaluated after illuminating the stalactites using exclusively white

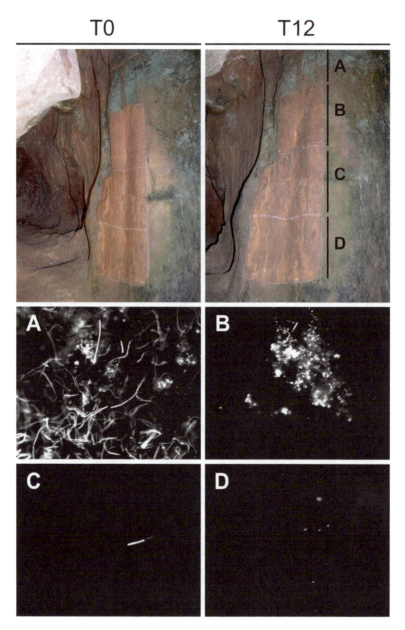

Fig. 4.8 Colonisation of cyanobacteria (*Scytonema julianum*, *Nostoc punctiforme*, and *Gloeocapsopsis magma*) in a stalactite, Collbato Cave, Spain, after cleaning treatments and green lighting for 12 months. Quadrant A is a control. Quadrant B was cleaned using 70% alcohol/ distilled water. Quadrants C and D were treated with benzalkonium chloride (1‰) and Preventol R50 (2%), respectively. Pigment fluorescence was used as an indicator of photosynthetic activity. Note the near absence of fluorescence after biocide treatments

light lamps in one of the test zones and green light lamps (Roldán et al. 2006) in the other zone.

After 1 year of treatment, the stalactite illuminated with white light was recolonised by phototrophic microorganisms, proving that none of the treatments was fully effective (Akatova et al. 2009). Increased biocide concentration (quadrant D) led to a progressive reduction but not a total inhibition of biofilm development. Roldán et al. (2006) proved that green lighting could prevent the growth of photosynthetic organisms, except for those capable of modifying accessory pigments. However, even biofilms composed of the chromatic adaptable phycoerythrin-containing *Gloeothece membranacea* had lower photosynthetic pigment biovolume, smaller thylakoid regions, and a weaker fluorescence intensity under green light than under white light, all of which are signs of retarded growth. These data and those from the Salpetre stalactite exposed to green light suggested that green lighting of show caves was a useful treatment for preventing photosynthetic biofilm growth in artificially illuminated works of art.

Albertano et al. (2005) monitored biofilm development using monochromatic lamps for laboratory tests and in situ experiments inside the Catacombs of St. Callistus and Domitilla in Rome (Italy). Three months of exposure to the green and blue monochromatic lights resulted in an effective reduction in cyanobacterial development. However, under blue light, the pH shift and the amount of soluble calcium within biofilms were lower than under green light, thus confirming that the poor development of cyanobacteria in these conditions resulted in a decreased mobilisation of the element from the calcareous substratum.

To avoid the use of bioremediation practices, which always have uncertain results, it is better to rely on the preventive conservation and the early detection of microbial outbreaks. To this aim, Porca et al. (2011) have proposed the use of an index of fungal hazard in show caves. This index is based on data on the concentration of fungal spores in the cave's air, knowledge of the cave's history and management, and a detailed survey of the different halls of the caves. The index classifies caves into five risk categories: category 1 designates a cave without a fungal problem, category 2 is a warning sign for caves, category 3 includes caves threatened by fungi, category 4 is assigned to caves already affected by fungi, and category 5 designates caves with an irreversible ecological disturbance.

Conclusions

Humans are the main threat for the conservation of caves. When visits cannot be prohibited, the maintenance of an exhaustive monitoring of possible changes occurring in the cave environment, particularly with regard to microclimate or new microbial colonisations, is necessary to control an outbreak in its initial stages. The creation of a Microbial Observatory is recommended in show caves subjected to high anthropogenic impacts. Frequent surveys of all cave elements can provide warnings for potentially dangerous situations. This survey work can be performed by guides and cave staff; these individuals should be in regular contact with microbiologists, who can identify the origin of the outbreaks and propose the most appropriate treatments.

Acknowledgments This work was supported by the Spanish Ministry of Science and Innovation, "Programa de investigación en tecnologías para la valoración y conservación del patrimonio cultural", TCP CSD2007-00058. Thanks go to Mr. Manuel Aragón, Nerva, Spain, for Figs. 4.3 and 4.4.

References

Akatova E, Roldan M, Hernandez-Marine M et al (2009) On the efficiency of biocides and cleaning treatments in restoration works of subterranean monuments. In: Science and cultural heritage in the Mediterranean area. Regione Siciliana, Palermo, pp 316–322

Albertano P, Bruno L, Bellezza S (2005) New strategies for the monitoring and control of cyanobacterial films on valuable lithic faces. Plant Biosyst 139:311–322

Bastian F, Alabouvette C, Saiz-Jimenez C (2009a) Bacteria and free-living amoeba in Lascaux Cave. Res Microbiol 160:38–40

Bastian F, Alabouvette C, Saiz-Jimenez C (2009b) Impact of biocide treatments on the bacterial communities of the Lascaux Cave. Naturwissenschaften 96:863–868

Bastian F, Alabouvette C, Saiz-Jimenez C (2009c) The impact of arthropods on fungal community structure in Lascaux Cave. J Appl Microbiol 106:1456–1462

Bastian F, Jurado V, Novakova A et al (2010) The microbiology of the Lascaux Cave. Microbiology 156:644–652

Bednarik RG (2007) Rock art science. The scientific study of palaeoart. Aryan Books International, New Delhi

Boston PJ, Northup DE, Lavoie KH (2006) Protecting microbial habitats. Preserving the unseen. In: Hildreth-Werker V, Werker J (eds) Cave conservation and restoration. National Speleological Society, Huntsville, AL, pp 61–82

Bretschneider T, Zocher G, Unger M et al (2012) A ketosynthase homolog uses malonyl units to form esters in cervimycin biosynthesis. Nat Chem Biol 8:154–161

Chalmin E, Menu M, Vingnaud C (2003) Analysis of rock art painting and technology of Palaeolithic painters. Meas Sci Technol 14:1590–1597

Clottes J, Courtin J, Valladas H et al (1992) La grotte Cosquer datée. B Soc Prehist France 89:230–234

Debbab A, Aly AH, Lin WH et al (2010) Bioactive compounds from marine bacteria and fungi. Microb Biotechnol 3:544–563

Dupont J, Jacquet C, Dennetiere B et al (2007) Invasion of the French Paleolithic painted cave of Lascaux by members of the *Fusarium solani* species complex. Mycologia 99:526–533

Elliott WR (2006) Biological dos and don'ts for cave restoration and conservation. In: Hildreth-Werker V, Werker J (eds) Cave conservation and restoration. National Speleological Society, Huntsville, AL, pp 33–46

Faimon J, Stelcl J, Kubesova S et al (2003) Environmentally acceptable effect of hydrogen peroxide on cave "lamp-flora", calcite speleothems and limestones. Environ Pollut 122:417–422

Fernandez-Cortes A, Cuezva S, Sanchez-Moral S et al (2011) Detection of human-induced environmental disturbances in a show cave. Environ Sci Pollut Res 18:1037–1045

Geneste J-M (2011) The major phases in the conservation of Lascaux Cave. In: Coye N (ed) Lascaux and preservation issues in Subterranean environments. Éditions de la Maison des Sciences de l'Homme, Paris, pp 51–71

Groth I, Saiz-Jimenez C (1999) Actinomycetes in hypogean environments. Geomicrobiol J 16:1–8

Groth I, Schumann P, Laiz L et al (2001) Geomicrobiological study of the Grotta dei Cervi, Porto Badisco, Italy. Geomicrobiol J 18:241–258

Hermosín B, Nováková A, Jurado V et al (2010) Observatorio microbiológico de cuevas: evaluación y control de comunidades fúngicas en cuevas sometidas al impacto de actividades turísticas. In: Durán JJ, Carrasco F (eds) Cuevas: Patrimonio, Naturaleza y Turismo. Asociación de Cuevas Turísticas, Madrid, pp 513–520

Jehlicka J, Vitek P, Edwards HGM et al (2009) Application of portable Raman instruments for fast and non-destructive detection of minerals on outcrops. Spectrochim Acta A 73:410–419

Jurado V, Sanchez-Moral S, Saiz-Jimenez C (2008) Entomogenous fungi and the conservation of the cultural heritage: a review. Int Biodeter Biodegr 62:325–330

Jurado V, Porca E, Cuezva S et al (2010) Fungal outbreak in a show cave. Sci Total Environ 408:3632–3638

Laiz L, Gonzalez JM, Saiz-Jimenez C (2003) Microbial communities in caves: ecology, physiology, and effects on paleolithic paintings. In: Koestler RJ et al (eds) Art, biology, and conservation: biodeterioration of works of art. The Metropolitan Museum of Art, New York, pp 210–225

Langsrud S, Sundheim G, Borgmann-Strahsen R (2003) Intrinsic and acquired resistance to quaternary ammonium compounds in food-related *Pseudomonas* spp. J Appl Microbiol 95:874–882

Lasheras JA, Sanchez-Moral S, Saiz-Jimenez C et al (2011) The conservation of Altamira Cave: a comparative perspective. In: Coye N (ed) Lascaux and preservation issues in Subterranean environments. Éditions de la Maison des Sciences de l'Homme, Paris, pp 169–182

Lavoie KH, Northup DE (2006) Bacteria as indicators of human impacts in caves. In: Rea GT (ed) 7th National cave and karst management symposium, proceedings. NCKMS Steering Committee, Albany, NY, pp 40–47

Lee NM, Meisinger DB, Aubrecht R et al (2012) Caves and karst environments. In: Bell EM (ed) Life at extremes: environments, organisms and strategies for survival. CAB International, Wallingford, pp 320–344

Malaurent P, Lacanette D, Brunet J et al (2011) Climatology of the subterranean environment at Lascaux: from a global study to the microclimatology of the cave walls. In: Coye N (ed) Lascaux and preservation issues in Subterranean environments. Éditions de la Maison des Sciences de l'Homme, Paris, pp 121–142

Martin-Sanchez PM, Sanchez-Cortes S, Lopez-Tobar E et al (2012a) The nature of black stains in Lascaux Cave, France, as revealed by surface-enhanced Raman spectroscopy. J Raman Spectrosc 43:464–467

Martin-Sanchez PM, Nováková A, Bastian F et al (2012b) The use of biocides for the control of fungal outbreaks in subterranean environments: the case of the Lascaux Cave in France. Environ Sci Technol 46:3762–3770

Mulec J, Kosi G (2009) Lampenflora algae and methods of growth control. J Cave Karst Stud 71:109–115

Northup DE (2009) Cave microbial communities: is protection necessary and possible? In: White WB (ed) Proceedings of 15th international congress of speleology, vol 2, Kerrville, TX, pp 763–767

Pettit P (2008) Art and the Middle-to-Upper Paleolithic transition in Europe: comments on the archaeological arguments for an early Upper Paleolithic antiquity of the Grotte Chauvet art. J Hum Evol 55:908–917

Pettit P, Pike A (2007) Dating European Palaeolithic cave art: progress, prospects, problems. J Archaeol Method Th 14:27–47

Pifferi A, Campi G, Giacovazzo C et al (2009) A new portable XRD/XRF instrument for non-destructive analysis. Croat Chem Acta 82:449–454

Porca E, Jurado V, Martin-Sanchez PM et al (2011) Aerobiology: an ecological indicator for early detection and control of fungal outbreaks in caves. Ecol Indic 11:1594–1598

Roebroeks W, Sier MJ, Nielsen TK et al (2012) Use of red ochre by early Neandertals. Proc Natl Acad Sci USA 109:1889–1894

Roldán M, Oliva F, González Del Valle MA et al (2006) Does green light influence the fluorescence properties and structure of phototrophic biofilms. Appl Environ Microbiol 72:3026–3031

Saiz-Jimenez C (2001) Estudio de los procesos de alteración de las rocas y pinturas rupestres de la cueva de Doña Trinidad (Ardales, Málaga) y abrigo de los Letreros (Vélez-Blanco, Almería). Panel 1:86–91

Saiz-Jimenez C, Cuezva S, Jurado V et al (2011) Paleolithic art in peril: policy and science collide at Altamira Cave. Science 334:42–43

Schabereiter-Gurtner C, Saiz-Jimenez C, Piñar G et al (2002a) Altamira cave Paleolithic paintings harbor partly unknown bacterial communities. FEMS Microbiol Lett 211:7–11

Schabereiter-Gurtner C, Saiz-Jimenez C, Piñar G et al (2002b) Phylogenetic 16S rRNA analysis reveals the presence of complex and partly unknown bacterial communities in Tito Bustillo cave, Spain, and on its Palaeolithic paintings. Environ Microbiol 4:392–400

Schabereiter-Gurtner C, Saiz-Jimenez C, Piñar G et al (2004) Phylogenetic diversity of bacteria associated with Paleolithic paintings and surrounding rock walls in two Spanish caves (Llonin and La Garma). FEMS Microbiol Ecol 47:235–247

Spötl C, Mattey D (2012) Scientific drilling of speleothems—a technical note. Int J Speleol 41:29–34

Stomeo F, Portillo MC, Gonzalez JM (2009) Assessment of bacterial and fungal growth on natural substrates: consequences for preserving caves with prehistoric paintings. Curr Microbiol 59:321–325

Valladas H (2003) Direct radiocarbon dating of prehistoric cave paintings by accelerator mass spectrometry. Meas Sci Technol 14:1487–1492

Valladas H, Cachier H, Maurice P et al (1992) Direct radiocarbon dates for prehistoric paintings at the Altamira, El Castillo and Niaux caves. Nature 357:68–70

Züchner C (2003) Archaeological dating of rock art. Nothing but a subjective method? INORA 35:18–24

Chapter 5
Microbial Ecology: Caves as an Extreme Habitat

C. Riquelme Gabriel and Diana E. Northup

Introduction

When you enter a cave, as a human, you are immediately struck by what an alien environment you have entered, as the light fades, and cool, moist air surrounds you. Using artificial light to illuminate your journey through this rock-dominated environment, you may think that nothing lives here, but biologists and microbiologists have been discovering a wealth of life in these rock chambers beneath the Earth's surface. And, now, a new revolution is taking place in how we view caves—scientists are discovering that these environments are home to organisms that produce secondary metabolites that may be useful to humans. But, what shapes this production and the microorganisms that produce these compounds? That is the focus of this chapter. We'll start with some background on caves in general, move to the abiotic factors that characterize caves, providing selective pressures for microbial evolution, and then review what energy sources fuel microbial growth and existence in caves, look at why caves make such ideal laboratories, and end with the significance of studying microbial life in caves, including secondary metabolite production, geomicrobiology, and relevance to life detection on extraterrestrial bodies.

Cave Zonation and Types

Caves are strongly zonal environments (Howarth 1983b; Poulson and White 1969), with up to five zones recognized, each with a different community of organisms and defined on the basis of its physical environment (i.e. the amount of light, moisture,

C.R. Gabriel (✉)
CITA-A, Departamento de Ciências Agrárias, Universidade dos Açores, Angra do Heroísmo, Portugal

D.E. Northup
Biology Department, University of New Mexico, Albuquerque, NM 87131, USA

N. Cheeptham (ed.), *Cave Microbiomes: A Novel Resource for Drug Discovery*,
SpringerBriefs in Microbiology 1, DOI 10.1007/978-1-4614-5206-5_5,
© Naowarat Cheeptham 2013

air flow, gas concentration, and evaporative power of the air). These are the **entrance, twilight, transition, deep**, and **stagnant air zones**. The entrance zone includes the area where we observe mixing of the surface and subsurface communities. The twilight zone is the area with reduced light between the limit of vascular green plants and the region of total darkness. The transition zone is characterized by total darkness and a variable abiotic environment that includes changing humidity level, airflow, and potential evaporation rate. In the deep cave zone the air remains still and saturated with water vapor; the substrate remains moist; and the potential evaporation rate is negligible over time. In the stagnant air zone, not always present, we see more restriction of air exchange, leading to occasional stagnation of the atmosphere. This air stagnation can be due to microbial respiration and gas concentrations, particularly carbon dioxide. These factors may be highly dynamic within given caves; however, the zonation can provide a valuable model to understand cave ecology, including microbial ecology, and the evolution of cave species (Howarth 1981a, b; 1983b). The size and shape of the passages, including constrictions, and location, size, and aspect of entrances control the presence and distribution of these zones within caves (Howarth 1993).

Palmer (2007) notes that the classification of cave types is "informal and flexible." Two major ways to classify caves include (1) how the cave originates (e.g., by dissolution) and (2) the host rock type (e.g., limestone). In category one, how the cave originates, dissolution and volcanic activity are the major methods by which caves originate and the caves that scientists working on secondary metabolites are likely to encounter. In category two, host rock type, those of greatest interest include limestone, gypsum, lava (usually basalt), granite, sandstone, and possibly ice caves. Fig. 5.1 shows a solution cave formed in gypsum (upper left), a solution cave formed in limestone (upper right), a lava cave formed by volcanic activity (lower right), and the ice formations that are often seen in the spring in lava caves (lower left), which are present as semipermanent ice formations. In solution caves you can see the evidence the movement of water last left on the cave passages. Lava caves are usually formed in pahoehoe (lava with a smooth, ropy texture) lava flows and are mainly basalt.

Cave Distribution

The best documented cave systems are in countries where caving has been popular for many years. A list of longest and deepest caves of the world is maintained on the Internet at http://www.caverbob.com/wlong.htm and http://www.caverbob.com/wdeep.htm, respectively. We present a sampling of the variety of long, deep, and interesting caves of the world below.

United States

Cave and karst (the solutional landscape characterized by caves and sinkholes) abound in the United States, with over 50,000 identified cave systems (Barton 2007).

Fig. 5.1 Photos copyright Kenneth Ingham

Major karst areas are found in the Tennessee (more than 7,000 caves), Alabama, and Georgia (the TAG region). The largest known cave in the world is the Mammoth Cave System in Kentucky, with 628 km (390 mi) of surveyed passage. The USA also has four other of the ten longest caves in the world, including Jewel Cave (253 km, 157 mi) and Wind Cave (221 km, 137mi) in South Dakota, Lechuguilla Cave

(210 km, 130 mi) in New Mexico, and Fisher Ridge Cave System (184 km, 114 mi) in Kentucky (http://www.caverbob.com/wlong.htm). Almost the entire state of Florida is karst, while major portions of Texas, Missouri, West Virginia, Virginia, Kentucky, and New Mexico are karst and underlain by major caves, to name a few of the major caving areas (See the map of karst areas in the United States in Veni 2002). Major volcanic areas and lava caves are found in California, Hawai'i, Idaho, New Mexico, Oregon, and Washington states (Veni 2002). The deepest cave in the United States is Kazamura (1102 m, 3,614 ft), a lava cave in Hawai'i, while the deepest cave in the continental United States is Lechuguilla Cave at 489 m (1,604 ft).

Fossils and Paintings

Caves in the United States contain evidence of human occupation in the southeast, southwest, and California in particular (Willey et al. 2009). Traces of rock art paintings in caves have been observed also (M. Bilbo personal communication).

Europe

In Ukraine the Giant Gypsum Caves of the Carpathian Mountains include the Optimistic Cave with a total distance of about 232 km (144 mi), making it the largest cave in Europe and one of the largest in the world. The Hölloch Cave system in Switzerland includes the second largest cave in Europe with near 200 km (124.3 mi) length. The Schlossberg caves in Germany are Europe's largest sandstone caves, containing long corridors that connect their huge, multitiered sandstone rooms. Estonia and Croatia hold the deepest vertical shaft with 603 m (1,978 ft) in Vrtoglavica Cave and 553 m (1,814 ft) at Patkov Gušt, respectively. The deepest caves in Europe are Lamprechtsofen Vogelschacht Weg Schacht in Austria 1,632 m (5,354 ft), Gouffre Mirolda in France, with 1626 m (5335 ft), Reseau Jean Bernard 1602 m (5256 ft) in the French Alps, and Torca del Cerro del Cuevón in Spain 1589 m (5213 ft). Vrelo Une in Croatia is one of the deepest underwater caves in the world with a reached depth of 205 m (672 ft). Sintzi Spring is the deepest underwater cave in Greece with 126 m (413 ft). Drach Caves, situated in Mallorca, Spain contains one of the largest subterranean lakes in the world, called Martel Lake (http://www.caverbob.com/wlong.htm and http://www.caverbob.com/wdeep.htm).

There are many glacial caves within the mountains in the northern European countries, such as Iceland, and also in the mountainous ones. We can find the largest ice cave in the world in Austria, Eisriesenwelt with 42 km (26 mi). Djonovica is a glacial cave in Macedonia, which extends about 600 m (2,000 ft) underground. Sea caves are abundant in Scotland, including Fingal's Cave, which inspired a composition of the same name by the German composer Mendelssohn. Tectonic plate movements formed neotectonic caves in Sweden within the last eight thousand to ten thousand years, such as Torkulla Kyrka, Gillberga Gryt, and Bodagrottorna caves (http://www.showcaves.com/english/explain/Speleolog). Lava caves are found in Iceland (covering 10% of the country, formed during the last 12,000 years), in the Canary Islands in Spain (Tenerife's Cueva del Viento, the fifth longest lava tube in the world with 27,000 year of history), and in the Azores Archipelago in Portugal (Gruta das Torres is among the twenty longest lava tubes of the world) (http://www. caverbob.com/lava.htm).

Fossils and Paintings

The earliest European cave paintings date to the Aurignacian, some 32,000 years ago. Just in France and Spain, 350 caves with prehistoric paintings have been discovered. Saiz-Jimenez and his colleagues (e.g., Saiz-Jimenez et al. 2011 and references therein) have done extensive work on the impact of exotic microorganisms and other organisms on the amazing cave paintings of these two countries. Coinciding with the advent of the Neolithic period, the Lascaux and Altamira are very well-known caves because of their paintings with Magdalenian style. Caves with fossils are spread across Europe, e.g., Bear's Cave (Peştera Urşilor) in Romania, and Veternica Cave in northern Croatia where Neanderthal fossils were also found. In Germany, the Messel Pit Fossil Site produced the largest known collection of fossils from the Eocene Age, which occurred between 36 million and 57 million years ago. Kent's Cavern in England is a source of much information on Paleolithic humans.

Asia

Shuanghe Dongqun Cave in China (http://www.hongmeigui.net/bibliography.php) is Asia's largest cave (128 km length and 593 m height; 79,537 mi and 1,946 ft). The deepest cave is Voronya Cave in Abkhazia, with a depth of 2,191 m (7,188 ft). Abkhazia also holds the second and third deepest caves, Sarma and Illyuzia-Mezhonnogo-Snezhnaya 1,760 and 1,753 m (5,774; 5,751 ft) (http://www.caver bob.com/wlong.htm). Other record caves in Asia are in Malaysia with the world's largest chamber, Sarawak Chamber, located in Gua Nasib Bagus Cave and Deer Cave, harboring what was described as the largest cave passage in the world until 2009 when Son Doong cave, in Vietnam, was explored and measured. Nowadays there is an open debate about which passage is larger (Laumanns and Price, 2010). Ordinskaya Cave is the larger underwater cave in Russia, the second in Eurasia, and the world's greatest gypsum cave. The estimated cave extension is 4400 m (2.7 mi); 4000 m (2.5 mi) constitutes the underwater part. Lava tubes are found in South Korea (Bilemot Gul and Manjang Gul are the seventh and tenth longest lava tubes of the world), Middle-East countries as Jordan (Al-Fahda Cave, 923.5 m; 3030 feet) and Saudi Arabia (Hibashi Cave, 690 m; 2264 ft), and Japan (about 95 caves, the largest of which is Mitsuike- Ana at 2202 m; 1.37 mi) (http://www.caverbob. com/lava.htm).

Fossils and Paintings

The Russian Kapova Cave is known by its Paleolithic paintings (mammoths, rhinos, horses, and bison), but also because of the existence of the human remains and animal bones. Bhimbetka rock shelters and caves exhibit the earliest traces of human life in India. The oldest paintings date back to around 12,000 years. Animals like rhinoceros, tigers, wild buffalo, bears, antelopes, boars, lions, elephants, lizards, and various community activities are represented. Cave paintings can be seen all over Thailand, e.g, Pha Taem National Park and Khao Pla Ra with drawings aged between 3000 and 5000 years old showing an agrarian society. There are various cave sites in Malaysia that have rock paintings, the oldest located at Niah Caves with an age of 1200 years. In Indonesia, drawings with an estimated age of 5,000 years old can be found in Leang Leang Cave as well as Neolithic paintings with an approximate age of 11,000 years in Padah-Lin Cave, in Myanmar. Paleolithic paintings and fingerprints (6,000-year-old) can be seen in caves on Yabrai Mountain in north China's Inner Mongolia Autonomous Region.

Gua Song Terus in the Gunung Sewu karst of Java is said to be the oldest site of human cave habitation in Southeast Asia, and dates back to 120,000 years ago. Some of Japan's oldest human bone fragments (15,000 and 20,000 years old) were found in a cave on Ishigaki Island. The Lenggong valley is one of the most important areas for archaeology in Malaysia, with Kota Tampan caves with old artifacts of about 74,000 years and Perak Man with human remains (10,000–11,000 years old) as representative examples.

The overarching message is that no matter what part of the world you reside in, there are fantastic caves in which scientific research can be conducted. These caves vary in depth and length, both factors that affect the microbial composition found in the caves.

How does energy enter caves?

Energy in caves is critical and often scarce, but it does exist. Organic carbon to fuel heterotrophy enters caves in a variety of ways (Barton and Jurado 2007). If the cave has a stream or river running through it, as do many caves in Europe and the eastern United States, for example, organic carbon, including plant debris, is brought in through the movement of water through the cave. Many arid-land caves, such as those in the southwestern United States, often lack streams, but water seeps into caves from the surface, following fractures and pore spaces in the overlying rock. Some organic carbon is present in the rock walls of caves, having been entombed in the rock as the rock was formed. Air currents that circulate in caves, especially those with more than one entrance, bring various reduced carbon energy sources, as well as particulate matter. Visitors to caves, be they animals such as bats, raccoons, etc., or humans, bring organic carbon with them in terms of hair, skin cells, and excrement. Chemolithoautotrophy, on the other hand, is also a viable means of growth for microorganisms in caves. In the arid-land caves of New Mexico, we have documented the presence of reduced iron and manganese in the carbonate bedrock, as well as organic carbon and very small amounts of nitrogen and other needed trace elements (Spilde et al. 2005). Some caves, such as Cueva de Villa Luz in Tabasco, Mexico, have reduced gases such as hydrogen, hydrogen sulfide, carbon monoxide, etc., entering the cave through springs with inlets in the caves (Hose et al. 2000). These reduced gases fuel extensive microbial communities, while the small amount of reduced iron and manganese fuel microbial communities on a more limited scale (Northup et al. 2003).

How do caves differ from surface habitats?

Caves are different from surface environments in some fairly fundamental and not so fundamental ways, including the amount of light, energy, and nutrients, the degree of weathering that occurs, and the stability of several abiotic factors. Beyond the twilight zone, caves are completely aphotic, which strongly influences the microbial communities and rules out phototrophy as a primary energy source for communities. The products of photosynthesis do seep or wash into caves from the surface, providing some of the organic carbon for communities. Arid-land caves may lack streams that would carry major organic matter loads, while caves such as Mammoth Cave have major rivers running through them. One of the primary characteristics of karst landscapes is the sinking streams, which carry surface influences into the caves. The lack of light has undoubtedly influenced microbial adaptations in caves, and we hypothesize that these adaptations become more pronounced with depth. The lack of light also has probably influenced the loss of UV resistance observed in cave bacteria (Snider et al. 2009). We have much to learn yet about other adaptations to the subsurface and the differing abiotic conditions observed in caves.

Temperatures in caves are usually close to the mean annual surface temperature over the cave, with some notable exceptions shown by Frank Howarth (e.g., 1983a). In temperate regions, outside air temperature in summer remains above the cave temperature, and the warm summer air entering caves can carry in moisture. The reverse is true during winter, when cool air entering caves tends to dry passages as the air warms. Thus, the boundary between the transition and deep cave zones often changes seasonally, i.e., moving closer to entrances in summer and deeper into caves during winter. In Hawai'i, as in the tropics generally, the surface temperature falls below the cave temperature during most nights (the tropical winter effect). This causes a daily cycle of drying in passages exposed to in-coming air, especially those sloping downward from an entrance. In addition, the warmer ambient temperatures in tropical caves allow the air to hold more moisture and to exert greater evaporative power. Thus the transition zone is often more conspicuous and extensive in tropical caves than it is in temperate caves. New efforts to assess climate in carbonate caves are underway that may shed light on climatic influences on microbial distribution (Pflitsch et al. 2010). Temperatures in caves can affect the composition of microbial mats, but we have no data yet on the effect of temperature on secondary metabolite production. However, it's a reasonable presumption that some of the very cold caves (e.g., lava caves at higher elevations) may contain microorganisms that produce cold-adapted enzymes.

Depending on the cave, the connectivity (hydrologic and other) between the surface and subsurface can vary from no connectivity to abundant connectivity. A study by Sarbu et al. (1996) demonstrated that Movile Cave in Romania is completely isolated from surface influences. Generally, this is an exception and many connections to the surface exist. Pipan et al. (2006) have documented the entry of copepods into caves through infiltrating water and Simon et al. (2007) have investigated nutrient cycling between the surface and subsurface, in one of the few papers on nutrient cycling in caves.

Nutrients do enter from the surface, but nutrient conditions are still generally oligotrophic in nature in caves. Oligotrophy stresses cave organisms and may act as a selective pressure. Koch (1997) suggests that organisms in low nutrient environments grow at very slow rates and that cultivation studies using standard amounts of nutrients simply provoke death. Boston has pioneered innovative cultivation efforts in caves over the last fifteen years and has demonstrated the need to inoculate and incubate cultures in the caves, for extended periods on very lean media that matches the parent environments with particular attention to the metals present (e.g., Boston et al. 2001; Rusterholtz and Mallory 1994; Spilde et al. 2005). Comparing cave molecular phylogenies to closest relatives often provides essential clues to enable researchers to enrich for particular substrate use, metabolic capabilities, and metal oxidizing abilities. This has been true in spite of the fact that nearest relatives may have radically different properties from one another. Such techniques can increase the retrieval of cave-adapted microorganisms for testing of secondary metabolite production.

Cave Habitats

Caves and different cave types (e.g., carbonate versus lava caves) contain a variety of habitats within them that differ in amounts and types of energy inputs. If a cave has a substantial, or even modest, bat population, there is a huge input of organic carbon, nitrogen, and phosphorus through the accumulation of guano and dead bats, all of which will support heterotrophic microbial populations. Other cave habitats, such as pools of water in arid-land caves, may be extremely oligotrophic and support lower microbial population levels. Caves with reduced gas inputs, such as some of the sulfur-based caves, on the other hand, may have habitats that are ideal for supporting chemolithoautotrophic microbial communities. These habitats span a continuum of organic and inorganic energy sources that will be explored below. Guano is an organic deposit common in caves derived mainly from bat excrements, but also from other organisms such as cave and camel crickets.

Fig. 5.2 Stream in a Slovenian cave. Photo by Diana Northup

Carbonate systems with streams

Carbonate caves with streams that flow through them are more common in temperate and tropical regions. Fig. 5.2 shows a stream in a Slovenian cave, where large volumes of water flow through the caves, potentially bringing in (and removing) nutrients. Studies by Simon et al. (2003) showed that epilithic biofilms in cave streams take up organic matter (particulate and dissolved) from the surface detritus infiltrating into the caves. These biofilms, in turn, form the basis for other parts of the food web in stream caves.

Carbonate systems with no stream inputs

Several interesting microbial community studies have been conducted in carbonate caves that shed light on some of the microbial inhabitants of these caves (e.g., Barton et al. 2007). Some carbonate caves, such as Lechuguilla Cave in Carlsbad Caverns National Park, New Mexico, USA, also host ferromanganese deposits on their walls and evidence exists to suggest that microorganisms are able to scavenge reduced manganese from the wall rock, which microorganisms oxidize to provide energy for growth, leaving behind the oxide waste products (Northup et al. 2003; Spilde et al. 2005). Pools in carbonate caves in semiarid regions are often isolated and represent another cave habitat. Studies have been conducted of their microbial inhabitants, which reveal diverse microbial communities (Shabarova and Pernthaler 2010). These pools are often very oligotrophic and our studies have shown that the communities are completely different from one cave to another and within caves (Hughes et al. unpublished results). These findings suggest that utilizing multiple sampling sites for cultivating isolates for secondary metabolite testing would be very useful.

Guano communities

Feces from a variety of animals that visit or live in caves provide habitats that are rich in nutrients, especially nitrogen, carbon, and phosphorus. Cave roosting bats, especially summer occupiers of caves, provide deep deposits of bat guano, which are inhabited by many invertebrates and microorganisms (Lavoie and Northup 2009). Little or no work has been done on the secondary metabolite production by guano microbial residents.

Lava caves

Lava caves occur in many regions of the Earth, including, for example: Hawai'i, Washington, Oregon, New Mexico (USA), the Azorean (Portugal) and Canary Islands (Spain) in the Atlantic Ocean, Japan, Korea, Australia, and Iceland. Howarth (1973, 1981a, b) described the Hawaiian volcanic cave ecosystem, which provides key facets for the study of controls on microbial diversity. The main energy sources in Hawaiian lava tube ecosystems are tree roots, which penetrate the lava for several meters, organic matter, which washes in with percolating rainwater and is carried by air currents, and accidentals, which are surface and soil animals blundering into the cave. Both living and dead roots, which represent a key energy resource, are utilized by a variety of organisms. Furthermore, both rainwater and accidentals often use the same channels as roots to enter caves, so that root patches often provide food for a wide diversity of cave organisms. Snider (2010) has shown that roots in lava caves

Fig. 5.3 Pseudoscorpions in bat guano (upper left); bat fur dreadlocks in a lava cave (upper right); *Embaphion* beetle and guano moth on guano (lower right); and looking for invertebrates in bat guano. Photos copyright Kenneth Ingham

Fig 5.4 Roots entering a Hawaiian lava cave. Photo copyright Kenneth Ingham

Fig 5.5 Villa Luz Life: Green biofilms (upper left); fish that feed on microbes (upper right); snottites (middle); black stream sediments (lower right); suspended sulfur produces a milky colored stream that exits the cave (lower left). Photos copyright Kenneth Ingham

can act as a conduit for microorganisms from the rhizosphere, contributing nutrients to the cave environment. Iron, sulfur, and manganese in the lava may also be important energy sources for chemolithotrophy, particularly for microbial communities in deep cave zones. These energy sources are also observed in Azorean lava caves, which our group has studied extensively (e.g., Northup et al. 2011, 2012, and references therein).

Sulfur Cave Habitats

In the last two decades, we have discovered a wealth of microbial diversity associated with caves, or portions of caves, that are characterized as having inputs of reduced sulfur compounds (reviewed in Engel 2007) and are often dominated by chemolithoautotrophs. Caves with sulfur inputs have some characteristic microbial inhabitants associated with the sulfur deposits including species from the *Proteobacteria*, in particular. Because of the oxidation of sulfide to sulfuric acid, many of the microbial inhabitants are acidophiles. Cueva de Villa Luz is a prime example from which some of our work is derived (Hose et al. 2000). Villa Luz has large inputs of hydrogen sulfide, as well as carbon monoxide, hydrogen, methane, and sulfur dioxide. The cave has a variety of redox conditions that range from highly reducing to highly oxidizing (−350 to +500, respectively, as measured with an ORP meter). The reducing environment of the black deposits underlying the stream that runs through the cave contains sulfate reducing bacteria, as well as other groups. The infamous snottites have abundant *Gammaproteobacteria*, related to *Acidithiobacillus*, and other groups. The phlegmballs that line the neck of the springs bring hydrogen sulfide into the cave, contain methanogens and *Proteobacteria*.

Geomicrobiology of Caves

Caves provide a vast habitat for microbial communities and it is no longer a surprise that these microorganisms interact with the bedrock and that many important mineral transformations occurring in cave environments are influenced somehow by microorganisms. The evolution of the science of cave geomicrobiology is correlated to the progress of chemical and microbiological techniques. Early studies led to the detection of the mechanisms for the deposition of calcite and other minerals in caves and recent molecular phylogenetic studies show how diverse microbial communities in caves are. Combining these techniques with geochemical and microbiological techniques, we now have a better understanding of how microorganisms are involved in processes as varied as speleothem (i.e., secondary mineral formations) deposition and cavern enlargement (Barton and Northup 2007; Cañaveras et al. 2006; Engel et al. 2004). Microbial communities harbor microorganisms as varied as those that obtain energy directly from the oxidation of reduced inorganic compounds (chemolithotrophs), and microorganisms that obtain energy from the oxidation of reduced organic matter (chemoheterotrophs). Complex prokaryotic communities with bacteria and archaea have been recovered on speleothems and mineral surfaces (Ikner et al. 2007; Legatzki et al. 2011; Northup et al. 2011), and also fungi (Novakova 2009; Vaugham et al. 2011). Interactions (competitive or cooperative) could be an advantage for the interacting species in a community, promoting molecular evolution through adaptation and counteradaptation to cave environmental conditions (Paterson et al. 2010).

Cave environments are home to several types of reactions, including both geological constructive and destructive processes. The microbial processes in caves often involve redox reactions (Barton and Northup 2007; Sasowsky and Palmer 1994) that modify microenvironmental conditions. Biogenesis includes processes of construction of substrates, such as trapping and binding of detrital grains to a substrate and calcite precipitation in caves. Microbial involvement has been supported in the formation of pool fingers, stalactites/stalagmites, cave pisoliths, and moonmilk (Jones 2010 and references therein). Erosional processes of microbial activity, such as boring and dissolution promote the breakdown of rock substrate, recycling essential nutrients, such as carbon, nitrogen, sulfur, and phosphorus, which are of extreme importance in geochemical cycles.

Fig. 5.6 Pool fingers in Carlsbad Cavern, New Mexico. Photo copyright Kenneth Ingham

Geomicrobial studies on the degradation of pigments and underlying rock material of Paleolithic paintings in caves promoted by microorganisms (Portillo et al. 2008; Saiz-Jimenez et al. 2011) have been instrumental in shaping conservation policies of the art inside those caves in relation to touristic visitation. In lava caves, microorganisms have been implicated in the formation of secondary minerals based on features such as stromatolitic textures, microbial fossils, and distinctive isotopic signatures (Léveillé and Datta 2010 and references therein).

Microbial metabolism is considered to be linked to the deposition of minerals in caves related to carbonates (reviewed in Dupraz et al. 2009); the sulfur cycle (Engel et al. 2004) and nitrate-rich deposits, such as saltpeter (Hill 1981). Bacteria have also been implicated in the formation of cave deposits containing iron oxide (de los Ríos et al. 2011; Northup et al. 2011) and manganese oxide (Northup et al. 2003; Spilde et al. 2005). Furthermore, most phosphate and some halite, gypsum, and other secondary mineral deposits in karst terrains are considered to be organic in origin (Taboroši 2006 and references therein). In its degradation process, guano reacts with bedrock and cave deposits and forms over 100 secondary minerals (Onac and Forti 2011).

Biologically driven processes involving silica have been described in volcanic caves. This is the case for volcanic cavities of Korea where the development of silica coralloids and helictites appeared to be linked to the presence in the twilight zone of diatom colonies of the genus *Meolosira* (Kashima et al. 1987). Another process in which microorganisms could be involved is the precipitation of silica moonmilk, formed by the weathering of basaltic rocks (Forti 2005). In the Algar do

Fig. 5.7 Opal speleothems in Algar do Carvão. Photo copyright Kenneth Ingham

Carvão, a volcanic pit on Terceira Island (Fig. 5.7, Azores), one of the largest deposits of opal flowstone in the world, appears to have formed through the diagenesis of
the silica moonmilk (Onac and Forti 2011).

Many geomicrobiological studies are currently underway to develop a better
understanding of reaction mechanisms occurring inside caves. All biologically
driven processes are complex and proving biogenicity (i.e., microbial involvement)
is difficult because the mere presence of microorganisms does not prove biogenicity
and many of the biogenic processes may mimic inorganic processes.

How might these conditions contribute to novel enzymes and secondary metabolites?

In an extreme environment, such as caves, microorganisms must cooperate to change their environment in their favor (Rosenberg et al. 1977). The demanding conditions of life in caves, mainly due to the scarcity of energy input, and the diversity of microorganisms living there, might suggest the presence of mutualistic interactions to support community growth under such starved conditions (Barton and Jurado 2007; Barton et al. 2007). The complex nature of the organic carbon and inorganic energy sources makes it highly unlikely that one organism is capable of carrying out all the reactions necessary to support growth (Laiz et al. 1999). There seem to be different strategies to achieve cooperative behavior. First of all, microbial communities in caves are organized into biofilms. As will be discussed later in more detail, biofilms provide protection against other cave inhabitants or adverse physical conditions (Jefferson 2004; López et al. 2010), spatial proximity that enables the communication among "neighbors," and resistance to many antimicrobials, through persister cells expressing toxin–antitoxin systems (Lewis 2005). Phenotypically distinct subpopulations of individuals often colonize biofilms, and a coordinated reaction occurs when sensing local environments. These specialized cell types often arise because of differences in gene expression, but not in gene composition (Stewart and Franklin 2008; Jefferson 2004). Microbial syntrophy is a special case of cooperation between metabolically diverse microorganisms that depend on each other for degradation of a certain substrate, typically through transfer of one or more metabolic intermediate(s) between the partners (Schink and Stams 2001; Wintermute and Silver 2010). Gene expression should be finally tuned to adapt to external conditions within such diverse microbial communities.

Quorum sensing is one of the processes, whereby microorganisms communicate with each other using diffusible signal molecules called auto inducers. It has been shown in numerous species that quorum sensing controls the production of secreted substances such as exoproteases, surfactants, and antibiotics (Chatterjee et al. 1995; Eldar 2011; Kleerebezem et al. 1997; Nakano et al. 1991). In addition to quorum sensing molecules, a diversity of secondary metabolites, such as antibiotics, pigments, and siderophores, are used as informational cues to instigate a collective behavioral change to environmental challenges as well (Atkinson and Williams 2009; López et al. 2010; Rahid and Ahmed 2005). The release of these signals leads to a different behavior depending on which molecules are used. Using a linguistic metaphor, signal molecules could be the words spoken by microorganisms, each having a different meaning. This complex network of microbial communication systems involves a huge number of secondary metabolites and enzymes. Some of these metabolites have been analyzed and proven to be bioactive, antibacterial, antifungal, antiviral, anticancer, insecticidal, algicidal, and immunosuppressive. Microorganisms forming microbial mats in caves, and in other extreme environments, are especially valuable as a source of new bioactive compounds for different biotechnological applications (Dobresov et al. 2011 and references therein).

These pristine locations promote the existence of genes shaped by natural selection under harsh conditions, developing metabolic capabilities found nowhere else. The difficult accessibility to such environments has reduced human disturbances, contributing to the conservation of that invaluable genetic diversity. In fact, novel diversity has been routinely found across phyla in caves from Hawaiian and Azorean microbial mats with several sequences being less than 90% similarity to any bacterial genetic sequence in NCBI's GenBank database (Northup et al. 2012).

Nowadays, genomic molecular techniques facilitate the discovery and industrial development of microbial secondary metabolites with interest to humankind. There is, at present, a clear need for achieving both technology development and sustainability. The huge richness in genetic biodiversity in microorganisms has supported life on our planet for many hundreds of years and provides us a great number of useful compounds that increase our welfare. Caves are untapped locations where precious molecules are being synthesized by their host microorganisms.

What do we know about how active microbes are in the subsurface?

The challenge for microbiologists wishing to explore the subsurface world of caves is to tackle the microbial community ecosystems as they occur in caves to gain a complete picture of what is happening. We must simultaneously understand the environmental composition, distribution, and processes taking place, as well as the microbial population structure and how they interact with each other and with the environment to understand life in cave biofilms.

The majority of active bacteria in nature are attached to surfaces and, therefore, it is important to understand bacterial life attached to cave rock surfaces (Molin and Givskov 1999). Microbial mats are very heterogenic, contain a complex network of microbial communication systems, and are very highly structured. Growth differentiation, chemotaxis, and cell-to-cell signaling enable biofilm communities to organize structurally in response to the external conditions and the activities of the different biofilm members (Davies et al. 1998; Tolker-Nielsen and Molin 2000). Fruiting bodies, biofilm channels and pores, microcolonies, and other features have been described as common biofilm structures (Korber et al. 1996; Stoodley et al. 1997). Not all the cells are active in biofilms when different processes are taking place, so knowing the metabolic state of the members of the community is of utmost importance to know the microorganisms involved in each of the processes.

Are the microorganisms whose DNA we are studying from caves actually active, or do they exist in a dormant or nonviable form? This is a key question for those wishing to culture them and study their secondary metabolite production. Direct estimation of respiring cells in ferromanganese deposits from Lechuguilla and Spider caves in the Guadalupe Mountains in New Mexico, yielded the metabolically active fraction of the microbial community (Spilde et al. 2005). The different layers

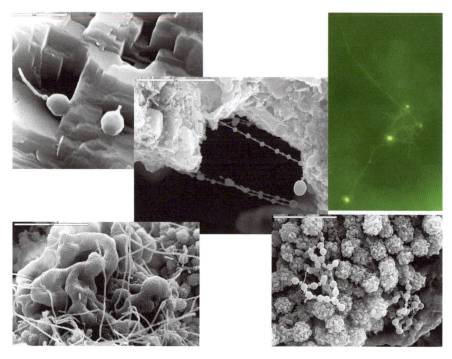

Fig. 5.8 Scanning electron micrograph (SEM) of beads on a string (upper left, middle). Acridine orange stained prostecate cells from Lechuguilla Cave (upper right); Microbial morphologies from Hawaiian lava caves (lower). Micrographs by Nortup, Spilde, and Rachel Schelble

of minerals and the rock underneath the ferromanganese deposit, termed "punk rock" (Hill 1987) were examined by epifluorescent microscopy after staining the cells with acridine orange and the metabolic (2-(*p*-iodophenyl)-3- (*p*-nitrophenyl)-5-phenyltetrazolium chloride (Northup 2002). Results showed active populations both in the ferromanganese deposits and in the "punk rock," which led to the hypothesis that bacteria were "mining" the substrate bedrock within the punk rock zone for reduced iron and manganese present in the carbonate bedrock, which is subsequently oxidized by iron- and manganese-oxidizing bacteria, and deposited at the cave air–rock interface (Northup 2002; Spilde et al. 2005).

Other approaches to determining the metabolically active microorganisms into the subsurface biofilms were performed using RNA-based molecular techniques in Altamira Cave, Spain. Gonzalez et al. reported members of the phylum Crenarchaeota as active components of microbial communities and potential participants in biodeteriorating processes (Gonzalez et al. 2006). Further analyses compared the total microbial community, based on DNA-based molecular techniques and the metabolically active members of the microbial community. The latter represented about 33% and 29% of the total microbial community in different colored colonies. The active transcripts recovered during this study belonged predominantly to the phylum *Proteobacteria*. The fraction of the total microbial community showing undetected metabolic activity represents a set of microorganisms with the potential of becoming metabolically active if the environmental conditions change (Portillo et al. 2008). Both anaerobic and aerobic bacteria were found to be active populations, which suggest a multilayered structure with anaerobic microenvironments within the depth of the microbial mat.

Implications for life on other planets

In addition to the value of studying secondary metabolite production by cave microorganisms, a strong motivation for the study of cave microorganisms, especially those in lava caves, is to enhance our ability to detect life on extraterrestrial bodies in the subsurface. Since similar atmospheric conditions were shared by Mars and Earth, with warmer temperatures and the presence of liquid water when life first evolved, the possibility of thriving life at the same time on the Red Planet became feasible (Baker et al. 1991; Beaty et al. 2005). Nowadays, the surface of Mars is exposed to UV radiation and ionizing radiation and temperatures below the freezing point of water, which makes the presence of life on the surface unlikely. Nevertheless, the subsurface of Mars is protected against such severe surface conditions, which led to the suggestion that life could exist there or traces of former life could be preserved (Boston et al. 2001; Izawa et al. 2010; Léveillé and Datta 2010). Lava tubes have been detected in the high-definition images available from the surface of Mars, and also by new technologies of remote detection (Cushing et al. 2007; Deák 2010). Because of the lower gravity on Mars, those lava tubes have to be much bigger than on Earth. Caves formed by mineral dissolution, may also be present beneath the surface because of the existence of sequences of evaporites in impact basins (Boston et al., 2006) and evidence of sedimentary deposits in the northern part of the planet, which could be a remnant of an ancient ocean (Baker et al., 1991; Clifford et al. 2001; Irwin and Schulze-Makuch 2010; Mouginot et al. 2012).

The search for life in Mars is focused on microorganisms rather than multicellular organisms due to the incompatibility of life, as we know it now, and the harsh conditions in the atmosphere of Mars. Since robotic missions to Mars are currently under development (Mars Exploration Joint Initiative signed in 2009 between US space agency, NASA, and European Space Agency, ESA), microbiological studies in Earth's caves can bring new insights in the search for life on Mars since researchers will need criteria with which select specific locations for analysis (Boston et al. 1992; Boston et al. 2001). Features in caves that look purely mineralogical actually contain extensive microbial communities. Molecular analyses of secondary mineral-like deposits found in Hawai'i, Azores, and New Mexico, including amorphous copper-silicate, iron-oxide formations, ooze-like coating, pink hexagons on basaltic glass and gold-colored deposits, revealed complex communities containing 14 phyla of bacteria across three locations (Northup et al. 2011). These deposits may also be fruitful grounds for culturing of isolates for testing of secondary metabolite production and provide completely new habitats in which to seek secondary metabolite producers. Metabolic capabilities of bacteria living in extreme environments (Kendrick and Kral 2006; Popa et al. 2012), will help in the development of new technologies needed to detect life on Mars (Chanover et al. 2011; Parro et al. 2011), or for the selection of pioneer microorganisms suitable to be introduced in Mars to modify its environment and make it habitable by ecopoiesis (Boston et al. 2009; Thomas et al. 2006).

Summary

To succeed in isolating microorganisms adapted to the subsurface that are potentially producing novel secondary metabolites, we need to really understand these native inhabitants of an environment that is totally foreign to us as humans. We need to envision the full range of factors that affect the microorganisms we wish to study, from the energy sources and trace nutrients that are available to the environmental conditions such as relative humidity and temperature to which the organisms are adapted. It's easy to realize that this is an aphotic environment in which phototrophy will not play a significant role, but to really understand the subtleties of the factors that have shaped the microorganisms over the long periods of time takes a keen observer with first-hand knowledge of what this alien environment is like. It also takes patience in cultivating the native inhabitants that grow slowly and extensive experimentation with different media techniques. But, the reward is potentially great because of the diversity of novel organisms that are being found in caves.

References

Atkinson S, Williams P (2009) Quorum sensing and social networking in the microbial world. J R Soc Interface 6:959–978
Baker V, Strom RG, Gulick VR, Kargel JS, Komatsu G, Kale VS (1991) Ancient oceans, ice sheets and hydrological cycle on Mars. Nature 352:589–594
Barton HA, Jurado V (2007) What's up down there: Microbial diversity in starved cave environments. Microbe 2:132–138
Barton HA, Northup DE (2007) Geomicrobiology in cave environments: Past, current and future perspectives. J Cave Karst Stud 69(1):163–178
Barton HA, Taylor NM, Kreate MP, Springer AC, Oehrle SA, Bertog JL (2007) The Impact of host rock geochemistry on bacterial community structure in oligotrophic cave environments. Int J Speleol 36:93–104
Beaty DW, Clifford SM, Borg LE, Catling DC, Craddock RA, Des Marais DJ, Farmer JD, Frey HM, Haberle RM, McKay CP, Newsom HE, Parker TJ, Segura T, Tanaka KL (2005) Key science questions from the second conference on early Mars: Geologic, hydrologic, and climatic evolution and the implications for life. Astrobiology 5:663–689
Boston PJ, Hose LD, Northup DE, Spilde MN (2006) The microbial communities of sulfur caves: A newly appreciated geologically driven system on Earth and potential model for Mars. In: R. Harmon (ed) Karst Geomorphology, Hydrology, & Geochemistry Geological Soc. Amer. Special Paper 404:331–344

Boston PJ, Ivanov MV, McKay CP (1992) On the possibility of chemosynthetic ecosystems in subsurface habitats on Mars. Icarus 95:300–308

Boston PJ, Spilde MN, Northup DE, Melim LA, Soroka DA, Kleina LG, Lavoie KH, Hose LD, Mallory LM, Dahm CN, Crossey LJ, Scheble RT (2001) Cave biosignature suites: microbes, minerals and Mars. Astrobiology 1:25–55

Boston PJ, Todd P, van de Kamp JL, Northup DE, Spilde MN (2009) Mars simulation challenge experiments: Microorganisms from natural rock and cave communities. Gravit Space Biol 22:2

Cañaveras JC, Cuezva S, Sanchez-Moral S, Lario J, Laiz L, Gonzalez JM, Saiz-Jimenez C (2006) On the origin of fiber calcite crystals in moonmilk deposits. Naturwissenschaften 93:27–32

Chanover NJ, Glenar DA, Voelz DG, Xifeng X, Tawalbeh R, Boston PJ, Brinckerhoff WB, Mahaffy PR, Getty S, ten Kate I, McAdam A (2011) An AOTF-LDTOF spectrometer suite for in situ organic detection and characterization. Aerospace Conf, IEEE 1–13

Chatterjee A, Cui Y, Liu Y, Dumenyo CK, Chatterjee AK (1995) Inactivation of *rsmA* leads to overproduction of extracellular pectinases, cellulases, and proteases in *Erwinia carotovora* subsp. *carotovora* in the absence of the starvation/cell density-sensing signal, N-(3-oxohexanoyl)-L-homoserine lactone. Appl Environ Microbiol 61:1959–1967

Clifford SM, Parker TJ (2001) The evolution of the Martian hydrosphere; implications for the fate of a primordial ocean and the current state of the Northern Plains. Icarus 154:40–79

Cushing GE, Titus TN, Wynne JJ, Christensen PR (2007) THEMIS observes possible cave skylights on Mars. Geophys Res Lett 34:L17201

Davies DG, Parsek MR, Pearson JP, Iglewski BH, Costerton JW, Greenberg EP (1998) The involvement of cell-to-cell signals in the development of a bacterial biofilm. Science 280:295–298

de los Ríos A, Bustillo MA, Ascaso C, Carvalho MR (2011) Bioconstructions in ochreous speleothems from lava tubes on Terceira Island (Azores) Sed Geol 236:117–128

Deák M (2010) The methodology of finding lava tubes with the use of remote detection on Mars, on the example of a newly found cave. 41st Lunar and Planetary Science Conference, The Woodlands, Texas. LPI Contribution No. 1533, p.1507, 1–5 March, 2010

Dobretsov S, Abed RMM, Al Maskari SMS, Al Sabahi JN, Victor R (2011) Cyanobacterial mats from hot springs produce antimicrobial compounds and quorum-sensing inhibitors under natural conditions. J Appl Phycol 23:983–993

Dupraz C, Reid RP, Braissant O, Decho AW, Norman RS, Visscher PT (2009) Processes of carbonate precipitation in modern microbial mats. Earth-Sci Rev 96:141–162

Eldar A (2011) Social conflict drives the evolutionary divergence of quorum sensing. Proc Nat Acad Sci USA 108(33):13635–13640

Engel AS, Stern LA, Bennett PC (2004) Microbial contributions to cave formation: New insights into sulfuric acid speleogenesis. Geology 32:369–372

Engel AS (2007) Observations on the biodiversity of sulfidic karst habitats. J Cave Karst Stud 69:187–206

Forti P (2005) Genetic processes of cave minerals In volcanic environments: An overview. J Cave Karst Stud 67(1):3–13

Gonzalez JM, Portillo MC, Saiz-Jimenez C (2006) Metabolically active Crenarchaeota in Altamira Cave. Naturwissenschaften 93:42–45

Hill CA (1981) Origin of cave saltpeter. Nat Speleol Soc Bull 43(4):110–126

Hill CA (1987) Geology of Carlsbad Cavern and other caves in the Guadalupe Mountains, New Mexico and Texas. New Mex Bur Mines Min Res Bull 117, Socorro, NM

Hose LD, Palmer AN, Palmer MV, Northup DE, Boston PJ, DuChene HR (2000) Microbiology the geochemistry in a hydrogen-sulphide-rich karst environment. Chem Geol 169:399–423

Howarth FG (1973) The cavernicolous fauna of Hawaiian lava tubes, 1. Introduction Pac Insects 15(1):139–151

Howarth, FG (1981a) Lava tube ecosystem as a study site. pp. 222–230. In: D. Mueller-Dombois, K.W. Bridges, H.L. Carson (eds.) Island Ecosystems: Biological Organization in Selected Hawaiian Communities. US/IBP Synthesis Series. Vol. 15. Hutchinson Ross Publishing Co., PA

Howarth, FG (1981b) Community structure and niche differentiation in Hawaiian lava tubes. Chapter 7. pp. 318–336. IN: D. Mueller-Dombois, K.W. Bridges, H.L. Carson (eds.) Island Ecosystems: Biological Organization in Selected Hawaiian Communities. US/IBP Synthesis Series. Vol. 15. Hutchinson Ross Publishing Co., PA

Howarth FG (1983a) Bioclimatic and geologic factors governing the evolution and distribution of Hawaiian cave insects. Entomol Gen 8:17–26

Howarth FG (1983b) Ecology of cave arthropods. Ann Rev Entomol 28:365–389

Howarth FG (1993) High-stress subterranean habitats and evolutionary change in cave-inhabiting arthropods. Amer Nat 142:S65–S77

Ikner LA, Toomey RS, Nolan G, Neilson JW, Pryor BM, Maier RM (2007) Culturable microbial diversity and the impact of tourism in Kartchner Caverns, Arizona. Microb Ecol 53(1):30–42

Irwin LN, Schulze-Makuch D (2011) Cosmic biology: How life could evolve on other worlds. J Mason (ed), Springer Praxis Books in Popular Astronomy. Science + Business Media

Izawa MRM, Banerjee NR, Flemming RL, Bridge NJ, Schultz C (2010) Basaltic glass as a habitat for microbial life: implications for astrobiology and planetary exploration. Planet Space Sci 58:583–591

Jefferson KK (2004) FEMS Microbiol Lett 236:163–173

Jones B (2010) Microbes in caves: agents of calcite corrosion and precipitation. In: Pedley HM & Rogerson M (ed), Tufas and speleothems: unraveling the microbial and physical controls. Geological Society, Special Publications, London, 336(1):7–30

Kashima N, Irie T, Kinoshita N (1987) Diatom, contributors of coralloid speleothems, from Togawa-Sakaidani-Do cave in Miyazaki Prefecture, Central Kyushu, Japan. Int J Speleol 16:95–100

Kendrick MG, Kral TA (2006) Survival of methanogens during desiccation: Implications for life on Mars. Astrobiology 6(4):546–551

Kleerebezem M, Quadri LE, Kuipers OP, de Vos WM (1997) Quorum sensing by peptide pheromones and two-component signal-transduction systems in Gram-positive bacteria. Mol Microbiol 24:895–904

Koch AL (1997) Microbial physiology and ecology of slow growth. Microbiol Mol Biol Rev 61:305–318

Korber DR, Lawrence JR, Lappin-Scott HM, Costerton JW (1996) The formation of microcolonies and functional consortia within biofilms. In: Lappin-Scott HM, Costerton JW (eds) Microbial Biofilms. Cambridge University Press, Cambridge, pp 15–45

Laiz L, Groth I, Gonzalez I, Saiz-Jimenez C (1999) Microbiological study of the dripping water in Altamira Cave (Santillana del Mar, Spain). J Microbiol Meth 36:129–138

Laumanns M, Price L (2010) Atlas of the great caves and the karst of Southeast Asia. Michael Laumanns & Liz Price (eds). Berliner Höhlenkundliche Berichte, (BHB) volume 40–41

Lavoie KH, Northup DE (2009) Invertebrate colonization and deposition rates of guano in a man-made bat cave, the Chiroptorium, Texas USA. Int Cong Speleol Proc 2:1297–1301

Legatzki A, Ortiz M, Neilson JW, Dominguez S, Andersen GL, Toomey RS, Pryor BM, Pierson LS III, Maier RM (2011) Bacterial and archaeal community structure of two adjacent calcite speleothems in Kartchner Caverns, Arizona, USA. Geomicrobiol J 28(2):99–11

Léveillé RJ, Datta S (2010) Lava tubes and basaltic caves as astrobiological targets on Earth and Mars: A review. Planet Space Sci 58:592–598

Lewis K (2005) Persister cells and the riddle of biofilm survival. Biochemistry 70:267–274

López D, Vlamakis H, Kolter R (2010) Biofilms Cold Spring Harb Perspect Biol 2:a000398

Molin S, Givskov M (1999) Application of molecular tools for in situ monitoring of bacterial growth activity. Environ Microbiol 1(5):383–391

Mouginot J, Pommerol A, Beck P, Kofman W, Clifford SM (2012) Dielectric map of the Martian northern hemisphere and the nature of plain filling materials. Geophysical Research Letters 39:L02202

Nakano MM, Magnuson R, Myers AM, Curry J, Grossman AD, Zuber P (1991) srfA is an operon required for surfactin production, competence development, and efficient sporulation in Bacillus subtilis. J Bacteriol 173:1770–1778

Northup DE (2002) Geomicrobiology of Caves. The University of New Mexico, Albuquerque, NM, Dissertation

Northup DE, Barns SM, Yu LE, Spilde MN, Schelble RT, Dano KE, Crossey LJ, Connolly CA, Boston PJ, Natvig DO, Dahm CN (2003) Diverse microbial communities inhabiting ferromanganese deposits in Lechuguilla and Spider Caves Environ Microbiol 5:1071–1086

Northup DE, Hathaway JJM, Snider JR, Moya M, Garcia MG, Dapkevicius MLNE, Riquelme C, Stone FD, Spilde MN, Boston PJ (2012) Life In Earth's lava caves: Implications for life detection on other planets. In: "Life On Earth and Planets". Editors: Arnold Hanslmeier, Stephan Kempe and Joseph Seckbach. Publisher: Springer. In Press

Northup DE, Melim LA, Spilde MN, Hathaway JJM, Garcia MG, Moya M, Stone FD, Boston PJ, Dapkevicius MLNE, Riquelme C (2011) Microbial communities in volcanic lava caves: Implications for life detection on other planets. Astrobiology 11(7):601–618

Novakova A (2009) Microscopic fungi isolated from the Domica Cave system (Slovak Karst National Park, Slovakia): A review. Int J Speleol 38(1):71–82

Onac BP, Forti P (2011) Minerogenetic mechanisms occurring in the cave environment: An overview International Journal of Speleology 40(2):79–98

Palmer AN (2007) Cave geology. Cave Books, Dayton, OH, USA

Parro V, de Diego-Castilla G, Moreno-Paz M, Blanco Y, Cruz-Gil P, Rodríguez-Manfredi JA, Fernández-Remolar D, Gómez F, Gómez MJ, Rivas LA, Demergasso C, Echeverría A, Urtuvia VN, Ruiz-Bermejo M, García-Villadangos M, Postigo M, Sánchez-Román M, Chong-Díaz G, Gómez-Elvira J (2011) A Microbial oasis in the hypersaline Atacama subsurface discovered by a life detector chip: Implications for the search for life on Mars. Astrobiology 11(10):969–996

Paterson S, Vogwill T, Buckling A, Benmayor R, Spiers AJ, Thomson NR, Quail M, Smith F, Walker D, Libberton B, Fenton A, Hall N, Brockhurst MA (2010) Antagonistic coevolution accelerates molecular evolution. Nature 464:275–278

Pflitsch A, Wiles M, Horrocks R, Piasecki J, Ringeis J (2010) Dynamic climatologic processes of barometric cave systems using the example of Jewel Cave and Wind Cave in South Dakota, USA. Acta Carsologica 39(3):449–462

Pipan T, Christman MC, Culver DC (2006) Dynamics of epikarst communities: microgeographic pattern and environmental determinants of epikarst copepods in Organ Cave, West Virginia. Am Midland Nat 156:75–87

Popa R, Smith AR, Popa R, Boone J, Fisk M (2012) Olivine-respiring bacteria isolated from the rock-ice interface in a lava-tube cave, a Mars analog environment. Astrobiology 12(1):9–18

Portillo MC, Gonzalez JM, Saiz-Jimenez C (2008) Metabolically active microbial communities of yellow and grey colonizations on the walls of Altamira Cave, Spain. J Appl Microbiol 104:681–691

Poulson TL, White WB (1969) The cave environment. Science 165(3897):971–981

Rachid D, Ahmed B (2005) Effect of iron and growth inhibitors on siderophores production by Pseudomonas fluorescens. African Journal of Biotechnology 4(7):697–702

Rosenberg E, Keller KH, Dworkin M (1977) Cell density-dependent growth of *Myxococcus xanthus* on casein. J Bacteriol 129:770–777

Rusterholtz K, Mallory LM (1994) Density, activity and diversity of bacteria indigenous to a karstic aquifer. Microbiol Ecol 28:79–99

Saiz-Jimenez C, Cuezva C, Jurado V, Fernandez-Cortes A, Porca E, Benavente D, Cañaveras JC, Sanchez-Moral S (2011) Paleolithic art in peril: Policy and science collide at Altamira Cave. Science 334:42–43

Sarbu SM, Kane TC, Kinkel BF (1996) A chemoautotrophically based cave ecosystem. Science 272:1953–1955

Sasowsky ID, Palmer MV (1994) Breakthroughs in karst geomicrobiology and redox geochemistry. Karst Waters Institute Special Publications 1. Karst Waters Institute, Inc., Charles Town, WV, USA

Schink B, Stams AJM (2001) Syntrophism among prokaryotes. In: Dworkin M, Falkow S, Rosenberg E, Schleifer K-H, Stackebrandt E (eds) The prokaryotes: An evolving electronic resource for the microbiological community, 3rd edn. Springer-Verlag, New York

Shabarova T, Pernthaler J (2010) Karst pools in subsurface environments: Collectors of microbial diversity or temporary residence between habitat types: Environ Microbiol 12:1061–1074

Simon KS, Benfield EF, Macko SA (2003) Food web structure and the role of epilithic biofilms in cave streams. Ecology 84(9):2395–2406

Simon KS, Pipan T, Culver DC (2007) A conceptual model of the flow and distribution of organic carbon in caves. J Cave Karst Stud 69:279–284

Snider JR, Goin C, Miller R et al (2009) Ultraviolet radiation sensibility in cave bacteria: Evidence of adaptation to the subsurface? Int J Speleol 38(1):13–22

Snider JR (2010) Comparison of Microbial Communities on Roots. Ceilings and Floors of Two Lava Tube Caves in New Mexico. Unpublished Master's Thesis, Albuquerque, NM

Spilde MN, Northup DE, Boston PJ, Schelble RT, Dano KE, Crossey LJ, Dahm CN (2005) Geomicrobiology of cave ferromanganese deposits: A field and laboratory investigation Geomicrobiol J 22:99–116

Stewart PS, Franklin MJ (2008) Physiological heterogeneity in biofilms. Nat Rev Microbiol 6:199–210

Stoodley P, Boyle JD, Dodds I, Lappin-Scott HM (1997) Consensus model of biofilm structure. In: Biofilms: community interactions and control, pp. 1–9. Wimpenny JWT, Handley PS, Gilbert P, Lappin-Scott HM, Jones M. BioLine (Eds), Cardiff, UK

Taboroši D (2006) Biologically influenced carbonate speleothems. Geol Soc Am Spec Pap 404:307–317

Thomas DJ, Boling J, Boston PJ, Campbell KA, McSpadden T, McWilliams L, Todd P (2006) Extremophiles For ecopoiesis: Desirable traits for and survivability of pioneer Martian organisms. Gravit Space Biol 19(2):91–104

Tolker-Nielsen T, Molin S (2000) Spatial organization of microbial biofilm communities. Microb Ecol 40:75–84

Vaughan MJ, Maier RM, Pryor BM (2011) Fungal communities on speleothem surfaces in Kartchner Caverns, Arizona, USA. Int J Speleol 40(1):65–77

Veni G (2002) Revising the karst map of the United States. J Cave Karst Stud 64:45–50

Willey P, Watson PJ, Crothers G, Stolen J (2009) Holocene human footprints in North America. Ichnos—Int J Plant Animal Traces 16(1–2):70–75

Wintermute EH, Silver PA (2010) Dynamics in the mixed microbial concourse. Genes Dev 24:2603–2614

Chapter 6
Studies of Antibiotic Production by Cave Bacteria

Elizabeth T. Montano and Lory O. Henderson

Introduction

Caves are outstanding geomicrobiological, ecological, and evolutionary laboratories. They are relevant to life detection on extraterrestrial bodies because of the discovery of lava caves on several Solar System bodies (Boston et al. 1992, Léveillé and Datta 2009, Rasmussen et al. 2009). Karst landscape studies are of global importance because karst terrains, which contain caves, characterizes up to 25% of many landscapes and carries up to 40% of drinking water (Jones 1997). In addition to these core attributes of caves and karst landscapes, we are now discovering that the microorganisms that inhabit these ecosystems may produce secondary metabolites that can be of value to humans.

Caves are inhabited by a multitude of novel *Actinobacteria* (Groth et al. 1999; Northup et al. 2003, 2011), the phylum from which two-thirds of our natural antibiotics originate (Kieser et al. 2000). There is strong potential for discovering novel bacterial isolates, including *Actinobacteria* that exhibit antimicrobial activity in the depths of carbonate caves. Our studies document the antimicrobial activity of cave bacterial isolates and focus on abiotic and biotic factors existing within New Mexican caves that may influence the production of antibiotics by bacterial inhabitants. Here we present the **hypotheses** and results of studies by two members of the University of New Mexico/New Mexico Tech Subsurface Life in Mineral Environments (SLIME) Team. We hypothesize that **1)** increased human visitation and nutrient levels will decrease the number of isolates exhibiting antimicrobial activity among cave bacterial isolates and **2)** the lack of nutrients (i.e., increasing oligotrophy) at increasing depths will have an observable effect on the number of

E.T. Montano (✉)
Department of Biology, University of New Mexico, Albuquerque, USA

L.O. Henderson
Department of Biology, University of New Mexico, Albuquerque, USA

N. Cheeptham (ed.), *Cave Microbiomes: A Novel Resource for Drug Discovery*,
SpringerBriefs in Microbiology 1, DOI 10.1007/978-1-4614-5206-5_6,
© Naowarat Cheeptham 2013

culturable bacterial cave isolates capable of antibiosis. These hypotheses give rise to questions regarding the availability of nutrients, such as carbon and nitrogen, with increasing depth and human visitation. Understanding the ecological conditions (e.g., nutrient levels) under which bacteria produce antibiotics will allow for more successful identification and isolation of new antibiotic-producing bacteria, ultimately leading to new antibiotics to which bacteria are not yet resistant. The discovery of new antibiotics in these cave environments will also give rise to increased efforts in cave conservation as cave managers and visitors become aware of the value of cave microorganisms.

Abiotic and Biotic Factors of Carbonate Caves

Carbonate caves differ from one another in the variety of abiotic and biotic factors hypothesized to contribute to bacterial antibiotic synthesis. In particular, depth, microhabitats, human visitation and impact, and nutrient availability are especially interesting parameters when investigating this phenomenon. The aforementioned abiotic and biotic factors may contribute to the microorganisms inhabiting a particular niche within the caves, and overall production of antibiotics in nature by cave bacteria.

Depth of Carbonate Caves

Ecological parameters associated with caves such as depth below the surface and aphotic conditions, may be important in explaining the particular trends in the data we observe. In our study caves differ in depth from six meters (Backcountry Cave [BCC]) to 44 meters (Spider Cave [SP]), to 192 meters (Fort Stanton Cave [FS]), to 325 meters (Carlsbad Cavern [CC]), and to 489 meters (Lechuguilla Cave [Lech]). This range of depth allows us to test whether nutrients decrease with increasing depth below the surface, and whether we retrieve greater or fewer antimicrobial isolates as we culture isolates deeper within the caves.

Microenvironments and Nutrients

Caves vary in the habitats (i.e., microenvironments or niches) that they possess whose characteristics function as important ecological factors governing microbial cave life; these habitats may include speleothems, moonmilk, water, and rock, as well as soils that comprise decomposed wall rock or infiltrating soil from the surface, all of which are habitats from which we cultured our isolates (Fig. 6.1). Moonmilk is a special type of carbonate speleothem with a "cream cheese-like

Fig. 6.1 Various microenvironments in which cave isolates were cultured. The microenvironments are the following (starting from the top left in a clockwise manner: Rock (speleothem), pool of water surrounded by moonmilk, water percolating into the cave on a stalactite (speleothem), measuring the pH of dry moonmilk, another speleothem sampled within the cave. Published with kind permission of © K. Ingham 2012

consistency" that is formed through precipitation of calcium carbonate (Barton and Jurado 2007). Moonmilk can be calcitic or, can be hydromagnesite, or huntite (Moore and Sullivan 1997). The former is believed to be biogenic with the aid of microorganisms (e.g., Barton and Northup 2007; Baskar et al. 2011). Moonmilk has a history of being used as a topical wound treatment in the Middle Ages and may be home to many antibiotic producing bacteria (Moore and Sullivan 1997). Moonmilk can form on the walls and ceiling of the cave or may be present in the bottom of pools. These microenvironments (i.e., moonmilk, rock, water, and soils) are the places within the cave typically targeted for culturing potential antibiotic producing bacteria. The varying nutrient levels available in these microhabitats may also affect the number of isolates identified with antimicrobial activity. Both studies reported here tested the influence of levels of nutrients such as carbon and nitrogen on the number of antimicrobial isolates recovered from differing levels of micronutrients.

The nutrients available in the cave (see discussion in the chapter by Northup and Riquelme, this volume), and their levels, will shape the community of bacteria found in different niches in cave ecosystems. Studies by Barton and Jurado (2007) have determined three ways for energy to enter caves: 1) atmospheric gases (N_2, CO_2) and organic molecules of different types, 2) various chemical compounds dissolved in surface waters infiltrating into caves through the overlying soils, and 3) reduced forms of manganese, iron, and other metal ions that exist in the bedrock that forms the cave passages. Some energy also enters cave ecosystems through visitation by humans or animals. Humans and other animals shed hair and skin cells that contribute to added carbon, while small animals likely carry in vegetative debris from the surface. Bacteria in caves oxidize reduced sulfur, iron, or manganese, hydrogen and methane, mobilize inorganic phosphates and hydrolyze proteins, lipids, and other macromolecules secreted by other community members (Barton and Jurado 2007).

Temperature and Humidity

Relatively constant temperatures and nearly 100% humidity (RH) characterize many cave passages and are additional abiotic factors that might influence bacterial metabolic functions (see further discussion in Northup and Riquelme, this volume). Caves can be categorized by the climate associated with their geographic location and are further identified as tropical, semiarid, mediterranean, temperate, alpine, and arctic (Ford and Williams 2007). The arid and semiarid areas of the United States containing the caves included in this study are geographically located in the west (Schumann 2007). Moderate to severe floods and droughts, a cyclical pattern that directly correlates with the El Nino-Southern Oscillation characterize this climate (Schumann 2007). Due to the semiarid climate and elevation of this region, carbonate limestone caves typically range in temperature from 15 to 18°C. Because all five of our study caves were similar in temperature and RH, we did not test this factor's effect on antimicrobial isolate recovery.

Human Visitation and Impact

Human visitation is a biotic factor that could influence the occurrence of antibiotic producing bacteria in cave ecosystems. The fragility of the internal geologic components of caves makes them increasingly susceptible to anthropogenic disturbances (Fernandez-Cortez et al. 2011). One of the most apparent impacts of cave visitors in subterranean environments is the increase in CO_2 concentration, water vapor, and temperature (Hoyos et al. 1998; Fernandez-Cortez et al. 2011). Human impact by tourism introduces a variety of foreign materials such as hair, skin cells, sweat, skin oils, fibers from clothing, and dust particles that could potentially provide nutrients to microbes (Ikner et al. 2007). Increased nutrient availability due to the rise in organic matter introduced to the cave by humans could bring in new bacteria that are associated with humans, which could influence the microbial density and community composition given sufficient organic carbon increases (Salmon et al.1995; Lavoie and Northup 2005), potentially affecting the production of antibiotics by subterranean bacteria (Davelos 2004).

Field and Collection Sites of Parallel Studies

Field Sites

Four semiarid carbonate caves in Carlsbad Caverns National Park (CCNP), located southwest of Carlsbad, New Mexico (SE New Mexico, USA), where our laboratory has a long history of research effort and collecting permits, were chosen on the basis of their variation in age, depth, length, microhabitats, and human visitation. These caves include the Deep Zone of BCC, LHT in Carlsbad Cavern, Lech, and Spider Cave (SP). An additional site at Fort Stanton Cave, located near Capitan in southeastern New Mexico, was sampled particularly for its high level of human visitation. These caves differ in various abiotic and biotic factors that could impact secondary metabolic production by cave bacteria (Table 6.1).

Collection Sites for Hypothesis 1

Three caves were chosen to test this hypothesis based on the level of human impact present in each cave. BCC is closed to the public, and therefore has little human impact. However, Spider Cave and Fort Stanton Cave have weekly cave tours or informal visitation, respectively, while LHT has daily tours, indicating a higher human impact. Within each cave, areas were sampled that contained varying levels of moisture based on the evidence of water flow (characterized as dry, moist, and wet), soil, and moist rock walls. Ten sites were sampled per cave. Three replicate cultures were gathered within the same microhabitat at each of the ten sites in each cave.

Collection Sites for Hypothesis 2

To test the effect of increasing oligotrophy with depth, we sampled four caves that varied substantially in depth: BCC (6 m deep); Spider Cave (44 m), Lech (489 m), and Carlsbad Cavern (325 m). Sampling in BCC varied from 5 to 6 m of depth below the surface. Sampling in Spider Cave varied from 38 to 39 m of depth, while sampling in Lech was done at every 50 m of increasing depth below the surface

Table 6.1 Abiotic and Biotic Factors of Five NM Sampled Carbonate Caves. Samples included rock, cave soil, water, moonmilk, and ferromanganese deposits (FMD). A study by Polyak et al. 1998 established the age of the caves

Caves	Age (MY)	Sample Depth (m)	Length	Average Temperature (°C)	Microhabitats	Percent Carbon	Percent Nitrogen	Human Visitation
Spider Cave (SP)	6.0-11.3	38-39	195 m	17.6	Rock, cave soil	9.14	0.23	Moderate
Backcountry Cave (BCC)	6.0-11.3	1.5-6	23 m	15.8	Rock, cave soil	9.25	2.99	Low
Carlsbad Cavern: Left Hand Tunnel (LHT)	4.0-3.9	228	402 m	15.4	Moonmilk, water, rock	4.98	0.04	High
Fort Stanton Cave (FS)	0.5	192	25 km	14	Rock, cave soil	5.29	0.09	High
Lechuguilla (Lech)	5.7-6.0	150-300	209 km	20	Rock, FMD	-	-	Low

beginning with -150 m and ending at -300 m (Deeper regions will be sampled in the future). Sampling in Carlsbad Cavern was done in LHT passage, which is located 228 m below the surface and samples were taken within a range of 50 m of each other. LHT contains the greatest variety of microhabitats, which includes moon-milk, water, and rock (wall formations and ground pebbles). Because of the relative shortness of BCC, sampling was done every three meters, while sampling in Spider Cave was done approximately every 10 m. Each site was sampled three times for replication. The same triplicate sampling was followed in both Spider Cave and BCC, where there were nine sites each; rock and cave soil microhabitats were sampled in each. Across the four caves, we had depths of 1.5 to 300 m for testing our hypothesis. All samples were taken in the aphotic zone, except for entrance samples in BCC and Spider Cave.

Methods and Materials

Media

To increase the diversity and culturability of our isolates, both studies incorporated a variety of media types for onsite culturing within each of the four carbonate caves. These media included a low nutrient R2A (Difco™), ½ R2A enriched with carbonate rock (to mirror the nutrients available in the host rock), and Actinobacteria Isolation Agar (AIA) (Difco™), which is particularly effective for isolating the largest Phylum of antibiotic producing bacteria.

Culturing and Subculturing

Samples from the habitats above were obtained aseptically using sterile rayon-tipped swabs dipped in sterile deionized water prior to rubbing on the sampling surface (except for pool water, Fig. 6.2a). Swabs were then spread across the culture plate (Fig. 6.2b).

To maximize the potential for successful media colonization, cave cultures were incubated on site in the cave for 24 h. These cultures were subsequently incubated in the lab at 15 °C, a temperature that mimics the cave environment, in a dark incubator, and subcultured onto R2A to obtain pure isolates that were characterized morphologically (Lammert 2007).

Pure Isolate Characterization

Isolates from our studies were initially Gram stained to determine cell morphology and arrangement and to check for isolate purity. For the investigation of hypothesis 2,

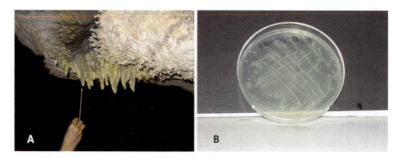

Fig. 6.2 a) Employing the sterile swab technique to a speleothem microenvironment. Published with kind permission of © K. Ingham 2012. b) The resulting bacterial culture plate derived from cave inoculation with kind permission of © K. Ingham 2012

Fig. 6.3 Examples of the range in the macromorphological characters belonging to bacterial colonies isolated from New Mexican carbonate caves and screened for antibacterial activity using the cross-streak assay. There are a variety of colors displayed here including pink, tan, white, and orange (upper left). The upper most colony is an example of a mucoidal formation (upper right). A successful streak isolation (bottom left) and an isolated colony of mucoidal formation and rose pigmentation (bottom right) are shown

we also recorded colony pigmentation to compare (1) isolates that exhibited antimicrobial activity to those that did not, (2) inhabitants within each of the sampled carbonate caves, and (3) the microenvironments from which isolates were obtained.

Nutrient Analyses

Soil/rock samples were collected from each sample site, in all caves for determination of organic carbon and nitrogen. Samples were collected in 50 cc Falcon tubes by scooping soil into the tube, or chipping rock into the tube. Samples were returned to the lab and stored at 4°C, until time of analysis. Soil/rock samples were then desiccated, ground to a fine powder, and analyzed. Percent organic carbon and nitrogen were determined by high temperature combustion. The resulting gases were eluted on a gas chromatography column and detected by thermal conductivity and integrated to yield carbon and nitrogen content (Pella, 1990a, 1990b). Analyses were performed on a ThermoQuest CE Instruments NC2100 Elemental Analyzer, ThermoQuest Italia S.P.A., Rodano, Italy.

For study 1, nutrient levels (i.e., low, medium, and high) were assessed based on the distribution of the percentages obtained. After graphically representing the data, break points were observed. These break points became the ranges that categorized low, medium, and high nutrient amounts. The nutrient ranges were the following: carbon (low: <2, medium: 2-9.999, high: ≥10) and nitrogen (low: <1, medium: 1-3.999, high: ≥4).

Antimicrobial Activity Assay

A modified version of the agar diffusion method (cross-streak assay) was used to test for antimicrobial activity among cave bacterial isolates. The initial cross-streak assay employed eight target bacteria: *Staphylococcus aureus* (ATCC 6538), *Klebsiella pneumoniae* (ATCC 13883) *Proteus vulgaris* (ATCC 13315), *Shigella flexneri* (ATCC 9199), *Streptococcus agalactiae* (ATCC 27956), *Streptococcus pyogenes* (ATCC 19615), and *Streptococcus pneumoniae* (ATCC 6303). Bacterial cultures isolated from the caves were inoculated in Nitrate Broth (Difco™) and incubated at cave temperature (15–20°C) for approximately 48 h or until visible growth was obtained. After the liquid (nitrate) incubation period, the cave cultures were inoculated (one vertical streak) onto ½ R2A medium. The cave bacterial isolate streaks were then grown at cave temperature for about 72 hours in order to obtain a thin line of growth. On the second day of the plate incubation, pathogenic bacteria were prepared in Plate Count Broth and grown for 12–18 h at 37°C. A thin Plate Count Agar (PCA) overlay (approximately 45°C) was then poured on top of the cave isolate streaks. A 1μL loopful of each of the target bacteria was horizontally inoculated on each previously vertically streaked cave isolate. The plates were incubated at 37°C

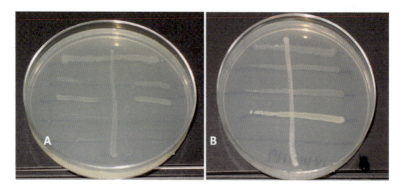

Fig. 6.4 a) Positive results of the cross-streak assay displaying inhibition of the second and third horizontally streaked target bacteria (i.e., pathogens). b) Negative results of the cross-streak assay displaying no inhibition among target bacteria

for 18–24 h, and then viewed for inhibition, which was defined as growth inhibition near the cave isolate streak. No inhibition was defined as no interruption of the target bacterial streak where it crossed the cave isolate streak (Fig. 6.4a-b). Gram-negative *Pseudomonas aeruginosa* ATCC 27853 never grew well and has no longer been used as a target bacterium. Gram-positive *S. pyogenes* ATCC 19615 and *Streptococcus were agalactiae* ATCC 27956 were also discontinued as targets because of their low rates of inhibition against the bacteria cultures from the caves in this study. This has subsequently narrowed the target bacteria to five.

Isolates Tested for Anitmicrobial Activity

Study 1:

The set of cultures taken from BCC, Spider Cave, and Fort Stanton Cave to test the effect of nutrient level and human visitation resulted in a total of 350 isolates. Out of 350 isolates, 335 were tested for antimicrobial activity; the remaining 15 isolates were lost to fungal contamination.

Study 2:

Cultures taken from Spider Cave, Carlsbad's LHT, Lech, and BCC resulted in 931 isolates to test the impact of increasing oligotrophy with depth below the surface. Out of 931 isolates, 690 were tested for antimicrobial activity; the remaining 241 isolates were lost to fungal contamination. Much of this loss resulted from isolates cultured from Carlsbad Cavern's LHT where culture plates contained no antifungal (Nystatin).

Results and Discussion

Hypothesis 1: Increased human visitation and higher nutrient levels will influence the number of isolates exhibiting antimicrobial activity

Bacteria that exhibited antimicrobial activity were isolated from both low and high human visitation carbonate caves. The results indicate that approximately 18.5% (62 out of 335) of the total isolates taken from BCC, Spider Cave, and Fort Stanton Cave are exhibiting antimicrobial activity. These caves had 21.1, 19.8, and 8.62 % of isolates demonstrating antimicrobial activity, respectively (Fig. 6.5). Future studies will include statistical analyses to determine whether the difference between antimicrobial activity in low (BCC) and high visitation caves (SP and FS) is significant. However, the difference observed among antimicrobial activity in these three caves could be attributed to the depths of each cave. Fort Stanton Cave is significantly deeper than BCC and Spider Cave. Fort Stanton Cave also has frequent flooding, which could significantly impact cave bacteria and their activities due to the influx of various organic matter or foreign bacteria or surface interactions. Thus, the higher levels of antimicrobial activity observed in the least visited cave, BCC, could be the result of lack of human impact, depth, or nutrient levels, or some synergy among these factors.

Antimicrobial activity as evidenced by zones of inhibition was observed most frequently when the cave isolates were tested against *S. aureus* (Gram-positive),

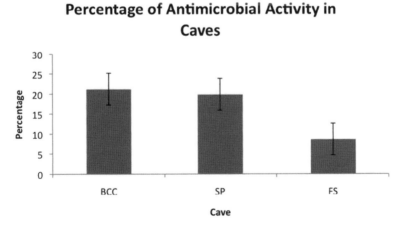

Fig. 6.5 Percentage of antimicrobial activity exhibited in all caves. The graph shows the percent of isolates that exhibited antimicrobial activity from each cave. Spider Cave is denoted as SP, Backcountry Cave as BCC, and Fort Stanton Cave as FS. Error bars represent the variance around the mean of one standard deviation

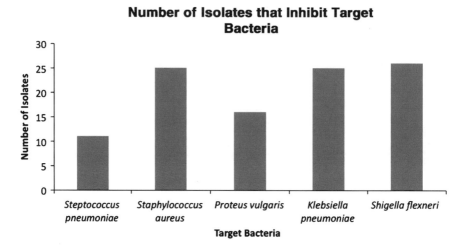

Fig. 6.6 Number of isolates that inhibit each target bacterium. The graph shows the number of positive isolates that produced antimicrobial activity from all caves that inhibit each of the five pathogens used in this study. Twenty-five (25), twenty-five (25), and twenty-six (26) isolates inhibit the growth of *Staphyloccocus aureus*, *Klebsiella pnuemoniae*, and *Shigella flexneri*, respectively

K. pneumoniae (Gram-negative), and *S. flexneri* (Gram-negative) (Fig. 6.6). The number of times each target bacterium was inhibited varied greatly, with some isolates inhibiting multiple target bacteria and some only inhibiting one.

The percentage of organic carbon represented in the caves shows the greatest variation, followed by nitrogen where the variance is low (Fig. 6.7). Overall, organic carbon levels were the highest of all three nutrients available throughout the cave environment, except in BCC.

A χ^2 Association Test was performed to determine whether there was a statistical difference among antimicrobial activity and various nutrient levels (i.e., low, medium, and high). The results indicate that there is no statistical difference in antimicrobial activity in areas of the caves characterized by low, medium, or high carbon ($\chi^2=0.558$, p$=0.757$) and nitrogen ($\chi^2=0.775$, p$=0.679$). This finding could be a result of a finding by Thomashaw et al. 2008. In surface terrestrial soil environments, it is the quantity and quality of the nutrients available and the ability of microorganisms to compete successfully for them that are the main determinants of microbial population size and perhaps microbial activity. It could be that the nutrients analyzed throughout the cave environment are similar in quality, which makes the level of nutrients insignificant in cave bacteria's ability to have microbial activity.

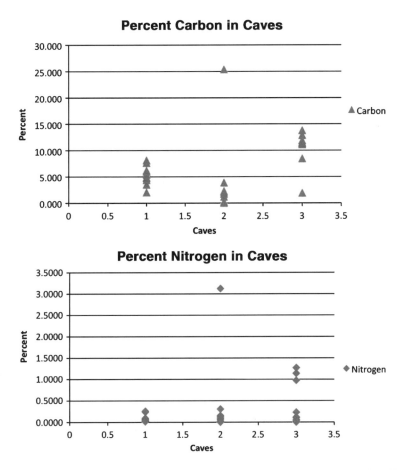

Fig. 6.7 Nutrient percentages in caves. The graphs show the percentage of nutrients found in all caves. The numbers on the horizontal axis refer to cave names: BCC (1), SP (2), and FS (3). The variation in the percentage of each nutrient decreases from carbon having the highest variance, to nitrogen with a lower variance

Hypothesis 2: The declining level of nutrients (i.e., increasing oligotrophy) at increasing depths will influence the number of isolates recovered that exhibit antimicrobial activity

The number of cultures that were successfully isolated and displayed antibiosis varied for each cave (Fig. 6.8). On average, 23.1% of all cave cultures produced an antimicrobial compound, a higher percentage than observed in study 1 caves, which were not as deep as some of the caves in study 2 caves. The number of times each target bacterium was inhibited varied greatly. The most frequently inhibited target bacteria across all four caves were Gram-negative; *K. pneumoniae* ATCC 13883, *P. vulgaris* ATCC 13315, and *S. flexneri* ATCC 9199 (Fig. 6.9). Those isolates that

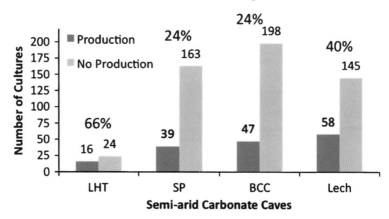

Fig. 6.8 The Left Hand Tunnel (LHT) passage in Carlsbad Cavern revealed 66% of the cultures were capable of antibiosis, whereas Spider (SP) and Backcountry Cave (BCC) were 24%, and Lechuguilla Cave (Lech) was 40%

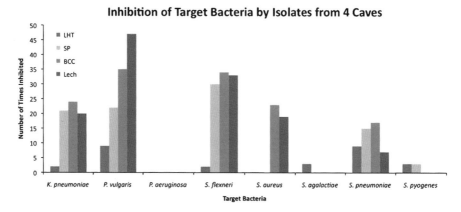

Fig. 6.9 The most frequently inhibited target bacteria in all four caves were Gram-negative; *Klebsiella pneumoniae* ATCC 13883, *Proteus vulgaris* ATCC 13315, and *Shigella flexneri* ATCC 9199. Culture isolates from Lechuguilla Cave were highly effective against *Proteus vulgaris*

exhibited antimicrobial activity against Gram-negative targets are of special interest because many of the antibiotic resistant pathogens today are Gram-negative bacteria (Arias and Murray 2009). Because of the range of response to target bacteria by cave isolates, a suite of target bacteria for testing appears to be a strong asset in detecting antimicrobial activity.

The subset of 164 bacterial cave isolates that produced visible signs of antimicrobial activity were assessed for whether or not there was a microhabitat, depth, or distance from the cave entrance wherein they most commonly exhibited antimicrobial activity against target bacteria in three (BCC, SP, LHT) of the four caves

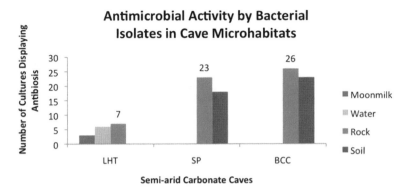

Fig. 6.10 Of the microhabitats sampled within three of the four sampled carbonate caves (Left Hand Tunnel (LHT) in Carlsbad Cavern, Spider (SP), and Backcountry Cave (BCC), the sediment (soil) yielded the most cultures capable of antibiosis. Moonmilk samples were only taken from LHT

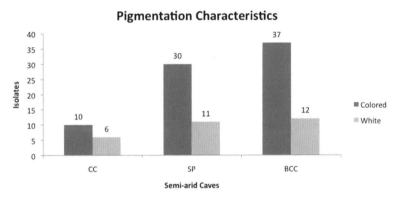

Fig. 6.11 Of the cultures capable of antibiosis that were sampled within three of the four sampled carbonate caves (Left Hand Tunnel (LHT) in Carlsbad Cavern, Spider (SP), and Backcountry Cave (BCC), colored morphology was more often observed than white

(Fig. 6.10). The most abundant isolates displaying antimicrobial activity were found within the rock microenvironment (Fig. 6.10), which includes many carbonate speleothems (Fig. 6.1). Surprisingly, isolates obtained from moonmilk had fewer isolates that exhibited antimicrobial activity. Because of the use of moonmilk in the Middle Ages as a wound treatment, we had hypothesized that this microhabitat would be a good source of antimicrobial activity isolates. This did not prove to be the case, but further testing of additional moonmilk sites is needed. Similar microenvironment data corresponding with Lech is currently being compiled and will soon be analyzed to investigate potential trends.

Colored isolates predominated over white colonies (67 of 106) in isolates exhibiting antimicrobial activity within three (BCC, SP, LHT) of the four carbonate caves investigated in this study (Fig. 6.11). One expectation was that cave-adapted bacteria would have lost their pigmentation in an effort to minimize energy expenditure.

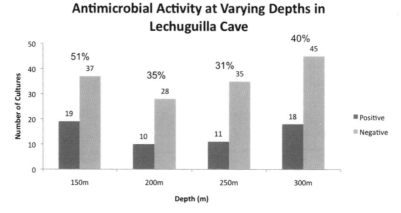

Fig. 6.12 The greatest numbers of isolates were successfully cultured from a depth of 300 m in Lechuguilla Cave. Of the isolates that displayed antibacterial properties, the highest percentages were at -150 m (51%) and -300 m (40%)

Whether our colored isolates are less cave-adapted than the white isolates will be tested in future investigations and will be extended to include Lech isolates.

In an effort try and correlate depth or distance from the cave entrance with bacteria capable of producing antimicrobial activity, these parameters were plotted against one another for each individual cave. Because of the geological structure of Lech, only depth from the surface was considered. Within Lech, the greatest percentages of successfully cultured bacterial isolates that displayed antibacterial properties were recovered from -150 m (51%) and -300 m (40%) (Fig. 6.12). It will be intriguing to determine whether or not samples taken above and below these depths will recover more or less antimicrobial activity producing isolates. Gathering samples at more depths will aid in this investigation. However, our findings to date may suggest that other factors are at play in determining where isolates with antimicrobial activity are recovered in Lech.

Spider Cave and BCC vary within a small range of depth, but provide information about how antibiosis changes with distance from the cave entrance. In Spider Cave, samples were taken at a depth of about -39 m; there is no trend in the number of isolates that produced antimicrobial compounds within the small range of this depth (Figs. 6.13, 6.14).

BCC is the shallowest of all of the study caves at only -6 m (Table 6.1) and hints at something else driving antibiosis. The most prolific producers were recovered from a depth of -1.59 m (Fig. 6.15), which corresponds to a distance of 16.75 m (Fig. 6.16) from the cave entrance, although there is no clear trend within the small range in depth for where the most producers can be found. What is most interesting about these data are the high numbers of producers recovered from the cave's entrance, which is located at -1.77 m below the surface (Figs. 6.15, 6.16). Because this area is nearest to the entrance and very shallow, nutrient inputs from the surface and vegetation surrounding the entrance may play a large part in this. BCC has the highest total percentages of both carbon and nitrogen compared to both Spider

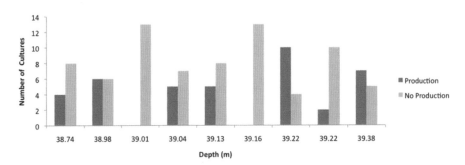

Fig. 6.13 There is no apparent trend in this narrow range in depths where bacteria capable of antibiosis are most abundant in Spider Cave. Although this displays the variance from cite to cite in the number of bacteria capable of antibiosis and where none were present

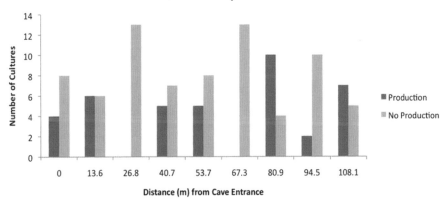

Fig. 6.14 Two distances 26.8 m and 67.3 m from the entrance of Spider Cave correspond to the depths (-39.01 m and -39.16 m) where no antimicrobial production was observed in cultured isolates

Cave and LHT (Table 6.1). Lech is expected to have the lowest total percentages of carbon and nitrogen; samples for analysis will be gathered for this in the future.

Our results suggest that nutrients, in some cases such as BCC, may correlate with higher numbers of isolates that exhibit antimicrobial activity. However, the preliminary results from Lech, a cave in which we expect the nutrient levels to be much lower than those found in all the other study caves based on previous carbon and nitrogen analyses by Northup (2002), this correlation is not substantiated. This suggests that additional factors, such as the substrate type, may have important

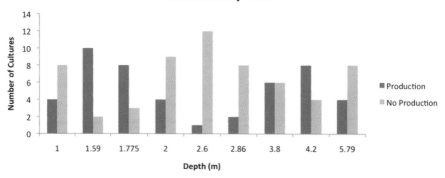

Fig. 6.15 There is no apparent trend in this narrow range in depths where bacteria capable of antibiosis are most abundant in Backcountry Cave. The most prolific producers were recovered from a depth of -1.59 m

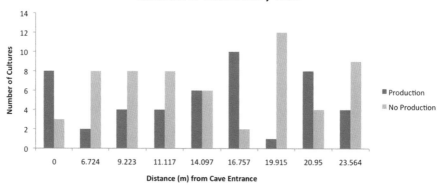

Fig. 6.16 The most prolific producers were recovered 16.7 m from the entrance of Backcountry Cave corresponding to a depth of -1.59 m. Note the elevated numbers of producers recovered from the entrance of the cave, where distance is zero, this is -1.77 m below the surface

influences on antimicrobial activity producing isolates. Future studies will sequence isolates that produce antimicrobial activity versus those that do not to investigate differences between these two groups by sampling site.

The amount of carbon decreases with depth below the surface; nitrogen also decreases with depth, but at least in three of our four study caves, it appears that once you reach a certain depth, nitrogen levels are relatively similar (Table 6.2). BCC, the shallowest sampled cave, had the highest carbon and nitrogen levels, with

Table 6.2 The table shows antimicrobial activity in isolates recovered from the caves used in this study, as well as abiotic factors (depth and average nutrient percentages) that could influence antimicrobial activity

Caves	Number Tested	Number Positive	Percent Antimicrobial Activity	Sample Depth	Percent Carbon	Percent Nitrogen
Backcountry Cave (BCC)	247	47	24	1.5-6 m	10.35	0.258
Spider Cave (SP)	202	39	24	38-39 m	7.81	0.032
Carlsbad Cavern: Left Hand Tunnel (LHT)	40	16	66	228 m	4.98	0.035
Lechuguilla Cave (Lech)	203	58	40	150-300 m	-	-

a corresponding level of bacterial antimicrobial activity of ~25%. Spider Cave, on the other hand, is deeper and had relatively the same percentage of antimicrobial activity isolates, but has a lower amount of both carbon and nitrogen. LHT in Carlsbad Cavern (CC) has the lowest abundance of both carbon and nitrogen of the three caves for which we have nutrient analyses; it is one of the two deepest caves sampled thus far, and has the highest number of isolates exhibiting antimicrobial activity. Thus, we support our hypothesis that increasing oligotrophy will lead to greater recovery of isolates that exhibit antimicrobial activity. However, in study 1, this difference was not observed, suggesting that further testing is necessary.

Conclusions

These studies document the presence of antimicrobial activity among cave bacteria isolates in carbonate caves, and also characterize select ecological conditions (nutrient levels and biotic and abiotic factors, such as human visitation and depth, respectively) under which isolates exhibiting antimicrobial activity occur. Based on the results obtained from both study 1 and 2, we can conclude that varying nutrient levels (i.e., low, medium, and high) within the caves do not influence antimicrobial activity in cave bacterial isolates, or do so in conjunction with other factors not yet identified. Further statistical analysis is needed to determine the impact of human visitation, but the similar percentages of antimicrobial activity seen in BCC (low visitation) and Spider Cave (high visitation) could indicate that human visitation does not have an effect on cave isolate antimicrobial activity. Results of study 2 allow us to conclude that increasing depth and increasing oligotrophy influence antimicrobial activity by cave bacterial isolates. Further investigation is needed to determine whether other factors (i.e., temperature, pH, age, etc., of caves), or the synergy among factors, could potentially govern the antimicrobial activity observed.

Significance

The genetic, chemical, and morphological characteristics belonging to the diverse microbial community inhabiting cave ecosystems, some of which exhibit antimicrobial activity, remain a black box. This research was established using the knowledge of previous studies, cave history, and cave geology. Our results are beginning to provide insight into the contents of the black box. These studies will also provide insight into the ecological processes that influence cave microbial isolates that exhibit antimicrobial activity in hopes of leading to conservation efforts put forth by cave managers. These experiments will provide information about the properties of cave-adapted bacteria and help determine their potential for antibiotic production. Results from these tests will identify a core set of isolates in which we will explore the nature of the antimicrobial activity documented, and the genes and secondary metabolite pathways associated with antibiotic production.

The ultimate goal of our studies is to determine the best habitats in which bacteria reside that possess antimicrobial activity against well-known pathogens, providing the basis for future research on novel antibiotics. Whether due to overuse or inadequate use, there is a great need to find new antibiotics to which bacteria are not yet resistant. When comparing genomes of surface and subsurface bacteria, we may find genes of ecological importance that are responsible for adaptation to the cave environment. The discovery of new antibiotics in carbonate cave environments will support more efforts in cave conservation as cave managers and visitors become aware of the value of cave microorganisms.

Furthermore, cave microbial communities may provide a valuable test bed for investigating the role that antibiotics play in natural settings. A study by Davies et al. (2006) suggested that antibiotic production in nature is to facilitate cell-to-cell communication, rather than as a "weapon." Such communication mechanisms could be useful to cave microbial communities. Because the availability of complex nutrients is limited in caves, there is evidence for community metabolism by a consortium of organisms working in concert (Boston, personal communication). Accumulation of metabolic products (for example, a secondary metabolite with antimicrobial properties) can create a mutualistic nutrient niche for supporting other microbes (Barton and Juardo 2007). Cave bacteria living syntrophically could utilize antibiotics to enhance nutrient acquisition for each other. If we find that antibiotics are produced in caves at low concentrations and more frequently in areas of exceedingly low nutrients, then we may contribute additional insight into this hypothesis of antibiotics as cell-to-cell communication. This finding provides further support for Peláez (2006), who suggests that natural products are the most promising source of novel antibiotics, and that cave environments have a great potential for new bioactive microbial metabolites.

Acknowledgements We thank the Cave Resources Office of Carlsbad Caverns National Park (Stan Allison and Dale Pate) for their support and collecting permits, and the Bureau of Land Management (Mike Bilbo and Jim Goodbar) for their support and collecting permits. This research is supported by the UNM IMSD (Initiatives to Maximize Student Diversity) program, NIH grant number: UNM-IMSD GM 060201, the American Society for Microbiology Undergraduate Research Fellowship, and the Cave Conservancy of the Virginias. Pat Cicero, Ara Kooser, and Wayne Walker provided invaluable field assistance. John Craig provided extremely helpful nutrient analyses and other laboratory support. Yvonne Bishop provided help with pathogen culturing. UNM undergraduates: Samantha Bear, Patty Murray, and Lariza Rosas and Sandia Prep High School students: Eddie Strach and Mark Holmen assisted in this research.

References

Arias CA, Murray BE (2009) Antibiotic-resistant bugs in the 21st century - a clinical super-challenge. N Engl J Med. doi:10.5495/wjcid.v1.i1.11

Barton HA, Juardo V (2007) What's up down there? Microbe-Am Soc Microbiol 2:132–138

Barton HA, Northup DE (2007) Geomicrobiology in cave environments: past, current and future perspectives. J Cave Karst Stud 69:163–178

Baskar S, Baskar R, Routh J (2011) Biogeneic evidences of moonmilk deposition in the Mawmluh Cave, Meghalaya, India. Geomicrobiol J 28:252–265. doi:10.1080/01490451.2010.494096

Boston PJ, Ivanov MV, McKay CP (1992) On the possibility of chemosynthetic ecosystems in subsurface habitats on Mars. Icarus. doi:10.1016/0019-1035(92)90045-9

Davelos AL, Kinkel LL, Samac DA (2004) Spatial variation in frequency and intensity of antibiotic interactions among *Streptomycetes* from prairie soil. Appl Environ Microbiol. doi:10.1128/AEM.70.2.1051-1058.2004

Davies J (2006) Are antibiotics naturally antibiotics? J Ind Microbiol Biotechnol. doi:10.1007/s10295-006-0112-5

Fernandez-Cortez A, Cuezva S, Sanchez-Moral S, Cañaveras JC, Porca E, Jurado V, Martin-Sanchez PM, Saiz-Jimenez C (2011) Detection of human-induced environmental disturbances in a show cave. Environ Sci and Pollut Res. doi:10.1007/s11356-011-0513-5

Ford D, Williams P (2007) Karst hydrology and geomorphology. Wiley and Sons, West Sussex

Groth I, Vetterman RB, Scheuetzte SP, Saiz-Jimenez C (1999) Actinomycetes in karstic caves of northern Spain Altamira, and Tito Bustillo. J Microbiol Methods. doi:10.1016/S0167-7012(99)00016-0

Hoyos M, Soler V, Cañaveras JC, Sanchez-Moral S, Sanz-Rubio E (1998) Microclimatic characterization of a karstic cave human impact on microenvironmental parameters of a prehistoric rock art cave (Candamo Cave northern Spain). Environ Geol 33:231–241

Ikner LA, Toomey RS, Nolan G, Neilson JW, Pryor BM, Maier RM (2007) Culturable microbial diversity and the impact of tourism in Kartchner Caverns. Arizona Microb Ecol. doi:10.1007/s00248-006-9135-8

Jones WK (1997) Karst hydrology atlas of West Virginia. Karst Waters Institute, Charles Town

Kieser T, Bibb MJ, Buttner MJ, Chater KF, Hopwood DA (2000) Practical *Streptomyces* genetics. John Innes Foundation, Norwich

Lammert JM (2007) Techniques in microbiology: A student handbook. Pearson Benjamin Cummings, San Francisco

Lavoie KH, Northup DE (2005) Bacteria as indicators of human impact in caves. Natl Cave and Karst Manag Sympos 40-47

Lertcanawanichakul M, Sawangnop S (2008) A comparison of two methods used for measuring the antagonistic activity of *Bacillus* species. Walailak J Sci & Tech 5:161–171

Léveillé RJ, Datta S (2009) Lava tubes and basaltic caves as astrobiological targets on Earth and Mars: a review. Planet Space Sci. doi:10.1016/ j.pss.2009.06.004

Moore GW, Sullivan N (1997) Speleology: caves and the cave environment. Cave Books, St. Louis

Northup DE (2002) Geomicrobiology of caves. University of New Mexico, Dissertation

Northup DE, Barns SM, Yu LE, Spilde MN, Schelble RT, Dano KE, Crossey LJ, Connolly CA, Boston PJ, Dahm CN (2003) Diverse microbial communities inhabiting ferromanganese deposits in Lechuguilla and Spider Caves. Environ Microbiol. doi:10.1046/j.1462-2920. 2003.00500.x

Northup DE, Melim LA, Spilde MN, Hathaway JJM, Garcia MG, Moya M, Stone FD, Boston PJ, Dapkevicius MENL, Riquelme C (2011) Lava cave microbial microcommunities within mats and secondary mineral deposits: Implications for life detection on other planets. Astrobiol. doi:10.1089/ast.2010.0562

Nurhidayu1 A, Ina-Salwany1 MA, Mohd Daud H, Harmin SA (2012) Isolation, screening and characterization of potential probiotics from farmed tiger grouper (*Epinephelus fuscoguttatus*). Afr J Microbiol Res. doi:10.5897/AJMR11.913

Palmer AN (2007) Cave geology. Cave Books, Dayton

Pella E (1990a) Elemental organic analysis part 1: historical developments. Am Lab 22:116–125

Pella E (1990b) Elemental organic analysis part 2: state of the art. Am Lab 22:28–32

Peláez F (2006) The historical delivery of antibiotics from microbial natural products-can history repeat? Biochem Pharmacol. doi:10.1016/j.bcp. 2005.10.010

Polyak VJ, McIntosh WC, Güven N, Provencio P (1998) Age and origin of Carlsbad Cavern and related caves from ^{40}Ar/^{39}Ar of alunite. Sci. doi:10.1126/science.279.5358.1919

Rasmussen B, Blake TS, Fletcher IR, Kilburn MR (2009) Evidence for microbial life in synsedimentary cavities from 2.75 Ga terrestrial environments. Geol. doi: 10.1130/G25300A.1

Salmon LG, Christoforou CS, Gerk TJ, Cass GR, Casuccio GS, Cooke GA, Leger M, Olmez I (1995) Source contributions to airborne particle deposition at the Yungang Grottoes China. Sci Total Environ 167:33–47

Schumann R (2007) The arid and semi-arid western United States. USGS Geological Survey: The National Assessment. US Dep Inter. http://esp.cr.usgs.gov/info/assessment/southwest.html Accessed 12 March 2012

Thomashaw LS, Bonsall RF, Weller DM (2008) Detection of antibiotics produced by soil and rhizophere microbes in situ. Springer, Heidelberg

Printed by Printforce, the Netherlands